Principles of Applied Statistics

Applied statistics is more than data analysis, but it is easy to lose sight of the big
picture. David Cox and Christl Donnelly draw on decades of scientific experience to
describe usable principles for the successful application of statistics, showing how
good statistical strategy shapes every stage of an investigation. As one advances from
research or policy questions, to study design, through modelling and interpretation,
and finally to meaningful conclusions, this book will be a valuable guide. Over 100
illustrations from a wide variety of real applications make the conceptual points
concrete, illuminating and deepening understanding. This book is essential reading for
anyone who makes extensive use of statistical methods in their work.

Principles of Applied Statistics

D. R. COX

Nuffield College, Oxford

CHRISTL A. DONNELLY

MRC Centre for Outbreak Analysis and Modelling,
Imperial College London

CAMBRIDGE
UNIVERSITY PRESS

CAMBRIDGE
UNIVERSITY PRESS

University Printing House, Cambridge CB2 8BS, United Kingdom

Cambridge University Press is part of the University of Cambridge.

It furthers the University's mission by disseminating knowledge in the pursuit of
education, learning and research at the highest international levels of excellence.

www.cambridge.org
Information on this title: www.cambridge.org/9781107013599

© D. R. Cox and C. A. Donnelly 2011

First published 2011

A catalogue record for this publication is available from the British Library

ISBN 978-1-107-01359-9 Hardback
ISBN 978-1-107-64445-8 Paperback

Contents

Preface

Statistical considerations arise in virtually all areas of science and technology and, beyond these, in issues of public and private policy and in everyday life. While the detailed methods used vary greatly in the level of elaboration involved and often in the way they are described, there is a unity of ideas which gives statistics as a subject both its intellectual challenge and its importance.

In this book we have aimed to discuss the ideas involved in applying statistical methods to advance knowledge and understanding. It is a book not on statistical methods as such but, rather, on how these methods are to be deployed. Nor is it a book on the mathematical theory of the methods or on the particular issue of how uncertainty is to be assessed, even though a special feature of many statistical analyses is that they are intended to address the uncertainties involved in drawing conclusions from often highly variable data.

We are writing partly for those working as applied statisticians, partly for subject-matter specialists using statistical ideas extensively in their work and partly for masters and doctoral students of statistics concerned with the relationship between the detailed methods and theory they are studying and the effective application of these ideas. Our aim is to emphasize how statistical ideas may be deployed fruitfully rather than to describe the details of statistical techniques.

Discussing these ideas without mentioning specific applications would drive the discussion into ineffective abstraction. An account of real investigations and data with a full discussion of the research questions involved, combined with a realistic account of the inevitable complications of most real studies, is not feasible. We have compromised by basing the discussion on *illustrations*, outline accounts of problems of design or analysis. Many are based on our direct experience; none is totally fictitious. Inevitably there is a concentration on particular fields of interest!

Where necessary we have assumed some knowledge of standard statistical methods such as least-squares regression. These parts can be skipped as appropriate.

The literature on many of the topics in the book is extensive. A limited number of suggestions for further reading are given at the end of most chapters. Some of the references are quite old but are included because we believe they retain their topicality.

We are grateful to the many colleagues in various fields with whom we have worked over the years, particularly Sir Roy Anderson, through whom we met in Oxford. It is a special pleasure to thank also Manoj Gambhir, Michelle Jackson, Helen Jenkins, Ted Liou, Giovanni Marchetti and Nancy Reid, who read a preliminary version and gave us very constructive advice and comments.

We are very grateful also to Diana Gillooly at Cambridge University Press for her encouragement and helpful advice over many aspects of the book.

1

Some general concepts

An ideal sequence is defined specifying the progression of an investigation from the conception of one or more research questions to the drawing of conclusions. The role of statistical analysis is outlined for design, measurement, analysis and interpretation.

1.1 Preliminaries

This short chapter gives a general account of the issues to be discussed in the book, namely those connected with situations in which appreciable unexplained and haphazard variation is present. We outline in idealized form the main phases of this kind of scientific investigation and the stages of statistical analysis likely to be needed.

It would be arid to attempt a precise definition of statistical analysis as contrasted with other forms of analysis. The need for statistical analysis typically arises from the presence of unexplained and haphazard variation. Such variability may be some combination of natural variability and measurement or other error. The former is potentially of intrinsic interest whereas the latter is in principle just a nuisance, although it may need careful consideration owing to its potential effect on the interpretation of results.

Illustration: Variability and error The fact that features of biological organisms vary between nominally similar individuals may, as in studies of inheritance, be a crucial part of the phenomenon being studied. That, say, repeated measurements of the height of the same individual vary erratically is not of intrinsic interest although it may under some circumstances need consideration. That measurements of the blood pressure of a subject, apparently with stable health, vary over a few minutes, hours or days typically arises from a combination of measurement error and natural variation; the latter part of the variation, but not the former, may be of direct interest for interpretation.

1

1.2 Components of investigation

It is often helpful to think of investigations as occurring in the following steps:

- formulation of research questions, or sometimes hypotheses;
- search for relevant data, often leading to
- design and implementation of investigations to obtain appropriate data;
- analysis of data; and
- interpretation of the results, that is, the translation of findings into a subject-matter context or into some appropriate decision.

This sequence is the basis on which the present book is organized.

It is, however, important to realize that the sequence is only a model of how investigations proceed, a model sometimes quite close to reality and sometimes highly idealized. For brevity we call it *the ideal sequence*.

The essence of our discussion will be on the achievement of individually secure investigations. These are studies which lead to unambiguous conclusions without, so far as is feasible, worrying caveats and external assumptions. Yet virtually all subject-matter issues are tackled sequentially, understanding often emerging from the synthesis of information of different types, and it is important in interpreting data to take account of all available information. While information from various studies that are all of a broadly similar form may be combined by careful statistical analysis, typically the important and challenging issue of synthesizing information of very different kinds, so crucial for understanding, has to be carried out informally.

Illustration: Synthesizing information (I)   One situation in which such a synthesis can be studied quantitatively occurs when several surrogate variables are available, all providing information about an unobserved variable of interest. An example is the use of tree ring measurements, pollen counts and borehole temperatures to reconstruct Northern Hemisphere temperature time series (Li *et al.*, 2010; McShane and Wyner, 2011).

Illustration: Synthesizing information (II)   The interpretation of a series of investigations of bovine tuberculosis hinged on a consistent synthesis of information from a randomized field trial, from ecological studies of wildlife behaviour and from genetic analysis of the pathogens isolated from cattle and wildlife sampled in the same areas.

Illustration: Synthesizing information (III)   The evidence that the Human Immunodeficiency Virus (HIV) is the causal agent for Acquired Immunodeficiency Syndrome (AIDS) comes from epidemiological studies of various high-risk groups, in particular haemophiliacs who received blood transfusions with contaminated products, as well as from laboratory work.

It is a truism that asking the 'right' question or questions is a crucial step to success in virtually all fields of research work. Both in investigations with a specific target and also in more fundamental investigations, especially those with a solid knowledge base, the formulation of very focused research questions at the start of a study may indeed be possible and desirable. Then the ideal sequence becomes close to reality, especially if the specific investigation can be completed reasonably quickly.

In other cases, however, the research questions of primary concern may emerge only as the study develops. Consequent reformulation of the detailed statistical model used for analysis, and of the translation of the research question into statistical form, usually causes no conceptual problem. Indeed, in some fields the refinement and modification of the research questions as the analysis proceeds is an essential part of the whole investigation. Major changes of focus, for example to the study of effects totally unanticipated at the start of the work, ideally need confirmation in supplementary investigations, however.

An extreme case of departure from the ideal sequence arises if new data, for example a large body of administrative data, become available and there is a perception that it must contain interesting information about something, but about exactly what is not entirely clear. The term 'data mining' is often used in such contexts. How much effort should be spent on such issues beyond the simple tabulation of frequencies and pair-wise dependencies must depend in part on the quality of the data. Simple descriptions of dependencies may, however, be very valuable in suggesting issues for detailed study.

While various standard and not-so-standard methods may be deployed to uncover possible interrelationships of interest, any conclusions are in most cases likely to be tentative and in need of independent confirmation. When the data are very extensive, precision estimates calculated from simple standard statistical methods are likely to underestimate error substantially owing to the neglect of hidden correlations. A large amount of data is in no way synonymous with a large amount of information. In some settings at least, if a modest amount of poor quality data is likely to be modestly misleading, an extremely large amount of poor quality data may be extremely misleading.

Illustration: Data mining as suggestive   Carpenter *et al.* (1997) analysed approximately 10^6 observations from UK cancer registrations. The data formed a 39×212 contingency table corresponding to 39 body sites and occupational coding into 212 categories. Although there is a clear research objective of detecting occupations leading to high cancer rates at specific body sites, the nature of the data precluded firm conclusions being drawn, making the investigation more in the nature of data mining. The limitations were that many occupations were missing, by no means necessarily at random, and multiple occupations and cancers at multiple sites were excluded. Crucial information on the numbers at risk in the various occupational categories was not available. A largely informal graphical approach to the data was used. Some previously well-established relationships were recovered; otherwise the conclusions were very tentative.

Illustration: Data mining or a fresh start   An industrial company that has specialized in making virtually the same product for many years may have very extensive records of routine tests made on the product as part of their quality control procedures. It may be reasonable to feel that important lessons could be learned from careful analysis of the data. But it is likely that the data have been recorded and the tests performed by different individuals using testing procedures that may have changed in important and possibly unrecorded ways and that many important changes affecting product quality are unrecorded. Extensive effort may be better directed at experimentation on the current system than at analyses of historical data.

Furthermore, while the sequence from question to answer set out above is in principle desirable, from the perspective of the individual investigator, and in particular the individual statistician, the actual sequence may be very different. For example, an individual research worker destined to analyse the data may enter an investigation only at the analysis phase. It will then be important to identify the key features of design and data collection actually employed, since these may have an important impact on the methods of analysis needed. It may be difficult to ascertain retrospectively aspects that were in fact critical but were not initially recognized as such. For example, departures from the design protocol of the study may have occurred and it very desirable to detect these in order to avoid misinterpretation.

1.3 Aspects of study design

Given a research question, a first step will be to consider whether data that may give at least a partial answer are already available. If not, one or more studies may need to be set up that are specifically aimed to answer the question.

The first considerations will then be the choice of material, that is, the individuals or contexts to study, and what measurements are appropriate. For a generic terminology we will refer to the individuals or contexts as *units of analysis*. We will discuss the criteria for measurements in more detail in Chapter 4. For studies of a new phenomenon it will usually be best to examine situations in which the phenomenon is likely to appear in the most striking form, even if this is in some sense artificial or not representative. This is in line with the well-known precept in mathematical research: study the issue in the simplest possible context that is not entirely trivial, and later generalize.

More detailed statistical considerations of design tend to focus on the precise configuration of data to be collected and on the scale of effort appropriate. We discuss these aspects in Chapters 2 and 3.

1.4 Relationship between design and analysis

The design and data collection phases of a study are intimately linked with the analysis. Statistical analysis should, and in some cases must, take account of unusual features of the earlier phases. Interpretation is the ultimate objective and so one objective of analysis is to get as close to incisive and justified subject-matter interpretation as is feasible.

Moreover, it is essential to be clear at the design stage broadly how the data are to be analysed. The amount of detail to be specified depends on the context. There are two reasons for requiring the prior consideration of analysis. One is that conclusions are publicly more convincing if established by methods set out in advance. An aspect that is in some ways more important, especially in studies with a long time frame, is that prior specification reduces the possibility that the data when obtained cannot be satisfactorily analysed. If, for quasi-legal reasons, for example to satisfy a regulatory agency, it is necessary to pre-specify the analysis in detail, it will be obligatory to follow and report that analysis but this should not preclude alternative analyses if these are clearly more appropriate in the light of the data actually obtained.

In some contexts it is reasonable not only to specify in advance the method of analysis in some detail but also to be hopeful that the proposed method will be satisfactory with at most minor modification. Past experience of similar investigations may well justify this.

In other situations, however, while the broad approach to analysis should be set out in advance, if only as an assurance that analysis and interpretation will be possible, it is unrealistic and indeed potentially dangerous to follow an initial plan unswervingly. Experience in collecting the data, and the data themselves, may suggest changes in specification such as a transformation of variables before detailed analysis. More importantly, it may be a crucial part of the analysis to clarify the research objectives; these should be guided, in part at least, by the initial phases of analysis. A distinction is sometimes drawn between pre-set analyses, called confirmatory, and exploratory analyses, but in many fields virtually all analyses have elements of both aspects.

Especially in major studies in which follow-up investigations are not feasible or will take a long time to complete, it will be wise to list the possible data configurations, likely or unlikely, that might arise and to check that data will be available for the interpretation of unanticipated effects.

Illustration: Explaining the unexpected In preliminary discussion of a study of hypertension, a group of cardiologists were unanimous that a certain intervention would lower blood pressure. When challenged as to their possible reaction if the data showed the opposite effect, their answer was that in five minutes a plausible explanation would be suggested and in ten minutes three different explanations. That is, even though there was an initial quite strong belief that a particular process would be operating, the possibility of alternative processes could and should not be totally excluded. This made it desirable, so far as was feasible, to collect data that might help clarify the situation should indeed blood pressure surprisingly increase following the intervention.

1.5 Experimental and observational studies

A crucial distinction is that we use the term *experiment* to mean a study in which all key elements are under the control of the investigator whereas a study is *observational* if, although the choice of individuals for study and of measurements to be obtained may be made by the investigator, key elements have to be accepted as they already exist and cannot be manipulated by the investigator. It often, however, aids the interpretation of an observational study to consider the question: what would have been done in a comparable experiment?

Illustration: Conflicting observational and experimental evidence
A number of observational studies, reviewed by Grady *et al.* (1992), suggested that women using hormone replacement therapy (HRT) for long periods of time had a lower coronary heart disease rate than apparently comparable control groups. In these studies the investigators chose the women to enter the investigation and the variables to be recorded but had no influence over whether any specific woman did or did not use HRT. In a randomized experiment (Writing group for Women's Health Initiative Investigators, 2002), women giving informed consent were assigned at random either to HRT or to an inactive control, the decision and its implementation being in principle concealed from the women and their treating doctor. After a period, this trial was stopped because of evidence of a possible *adverse* effect of HRT on total cardiovascular events and because of strong evidence of the absence of any useful overall beneficial effect. That is, the observational and experimental evidence were inconsistent with one another.

In its simplest terms the interpretation is as follows. The two groups of women compared in the observational studies may have been systematically different not only with respect to HRT use but also on a range of health-related features such as socio-economic status, education and general lifestyle, including eating habits. While some checks of comparability are possible, it remains the case that the clearly statistically significant difference in outcome between the two groups may be quite unconnected with HRT use.

By contrast, in the randomized trial the two groups differed only by the play of chance and by the fact that one group was allocated to HRT and the other to control. A clearly significant difference could confidently be taken as a consequence of the treatment allocation.

In an experiment conducted in a research laboratory the investigator could ensure that in all important respects the *only* difference between the groups being compared lay in HRT use versus no HRT use. The conclusion would thus be virtually unambiguous. In the more complicated environment of a clinical trial, however, especially one lasting an appreciable time, departures from the trial protocol might occur, such as a failure to take the allocated medication or the taking of supplementary medication, such departures being indirect consequences of the allocated treatment. Thus the primary comparison of outcomes in the clinical trial includes not only the direct effect of the medication but also its indirect effects. In this particular study, a modest but nonnegligible failure to comply with the study medication was reported but was judged not to modify the findings.

1.6 Principles of measurement

The primary requirements for measurements are:

- what is sometimes called *construct validity*, namely that the measurements do actually record the features of subject-matter concern;
- in particular that they record a number of different features sufficient to capture concisely the important aspects;
- that they are reliable, in the sense that a real or notional repetition of the measurement process gives reasonably reproducible results;
- that the cost of the measurements is commensurate with their importance; and
- that the measurement process does not appreciably distort the system under study.

We discuss measurements and most of the above points in more detail in Chapter 4. In particular, we note now that measurements can be classified by the structure of possible values (for example, binary or continuous) and, even more importantly, by their potential role in interpretation, for example as outcomes or as explanatory features.

The issue of dimensionality, especially that of the so-called outcome and response variables, depends strongly on the context.

Illustration: Summarizing multidimensional data   That the economic activity of a nation, the quality of life of an individual or the status of a complex organization such as a university can be wholly encapsulated in a single number such as a gross domestic product (GDP), a quality-adjusted life year (QUALY) or a league table ranking of the world's universities is patently absurd. In general, the description of complex multidimensional phenomena by a limited number of summary measures requires the careful specification of objectives. Pressure to produce one-dimensional summaries, to be resisted except for highly specific purposes, comes from the view that many situations can be explained in terms of the optimization of an appropriate one-dimensional criterion. This may be combined with the explicit or implicit assumption that utility can be measured in money terms.

1.7 Types and phases of analysis

A general principle, sounding superficial but difficult to implement, is that analyses should be as simple as possible, but no simpler. Some complication may be necessary to achieve precise formulation or to uncover the issue of interest from a confusing background or, somewhat less importantly, to obtain meaningful assessments of uncertainty.

Moreover, the method of analysis should so far as feasible be transparent. That is, it should be possible to follow the pathways from the data to the conclusions and in particular to see which aspects of the data have influenced key conclusions, especially any that are surprising. Black-box methods of analysis may to some extent be unavoidable in complex problems, but conclusions from them demand particularly careful checking and verification.

Four main phases of analysis are usually needed:

- data auditing and screening;
- preliminary analysis;
- formal analysis; and
- presentation of conclusions.

Data auditing and screening, which should take place as soon as possible after data collection, include inspection for anomalous values as well as for internal inconsistencies. Other relatively common sources of concern are sticking instruments, repeatedly returning the same value, for example zero rainfall, as well as the confusion of zero values and missing or irrelevant values. Sometimes, especially when extensive data are being collected in a novel context, formal auditing of the whole process of data collection and entry may be appropriate. Typically this will involve detailed study of all aspects of a sample of study individuals, and the application of ideas from sampling theory and industrial inspection may be valuable.

Methods of analysis are broadly either graphical or quantitative, the former being particularly helpful at the preliminary stages of analysis and in the presentation of conclusions. Typically, however, it will be desirable that in a final publication the key information is available also in numerical form, possibly as supplementary material. The reason is that reading data from graphs makes further analysis subject to (possibly appreciable) avoidable rounding error.

Graphical methods for handling large amounts of complex data, often studied under the name *visualization*, may require specialized software and will not be considered in detail here. For handling less complicated situations Section 5.4 suggests some simple rules, obvious but quite often ignored.

1.8 Formal analysis

Some methods of analysis may be described as *algorithmic*. That is to say, relationships within the data are recovered by a computer algorithm typically minimizing a plausible criterion. The choice of this criterion may not be based on any formal grounds and does not necessarily have any specific probabilistic properties or interpretation. Thus the method of least squares, probably the most widely used technique for fitting a parametric formula to empirical data, can be regarded purely algorithmically, as in effect a smoothing device; it gains some strength in that way. In most statistical settings, however, the method of least squares is formally justified by a probability model for the data.

We concentrate our discussions on analyses based on a formal probability model for the data, although certainly we do not exclude purely algorithmic methods, especially in the initial stages of the reduction of complex data.

1.9 Probability models

Most of our later discussion centres on analyses based on probability models for the data, leading, it is hoped, to greater subject-matter understanding. Some probability models are essentially descriptions of commonly occurring patterns of variability and lead to methods of analysis that are widely used across many fields of study. Their very generality suggests that in most cases they have no very specific subject-matter interpretation as a description of a detailed data-generating process. Other probability models are much more specific and are essentially probabilistic theories of the system under investigation. Sometimes elements of the second type of representation are introduced to amend a more descriptive model.

The choice of statistical model translates a subject-matter issue into a specific quantitative language, and the accuracy of that translation is crucial.

We return to these issues in depth in Chapter 6.

1.10 Prediction

Most of this book is devoted to the use of statistical methods to analyse and interpret data with the object of enhancing understanding. There is sometimes a somewhat different aim, that of empirical prediction. We take that to mean the prediction of as yet unobserved features of new study individuals, where the criterion of success is close agreement between the prediction and the ultimately realized value. Obvious examples are time series forecasting, for example of sales of a product or of the occurrence and amount of precipitation, and so on. In discussing these the interpretation of the parameters in any model fitted to the data is judged irrelevant, and the choice between equally well-fitting models may be based on convenience or cost.

Such examples are a special case of decision problems, in particular problems of a repetitive nature, such as in industrial inspection where each unit of production has to be accepted or not. The assessment of any prediction method has to be judged by its empirical success. In principle this should be based on success with data independent of those used to set up the prediction method. If the same data are used directly for both purposes, the assessment is likely to be misleadingly optimistic, quite possibly seriously so.

There are some broader issues involved. Many investigations have some form of prediction as an ultimate aim, for example whether, for a particular patient or patient group, the use of such-and-such a surgical procedure will improve survival and health-related quality of life. Yet the primary focus of discussion in the present book is on obtaining and understanding relevant data. The ultimate use of the conclusions from an investigation has to be borne in mind but typically will not be the immediate focus of the analysis.

Even in situations with a clear predictive objective, the question may arise whether the direct study of predictions should be preceded by a more analytical investigation of the usefulness of the latter. Is it better for short-term economic forecasting to be based on elaborate models relying on possibly suspect economic theory or directly on simple extrapolation?

Somewhat analogously, it may be better to base short-term weather forecasting on empirical extrapolation whereas longer term forecasting demands the well-established laws of physics.

An important and perhaps sometimes underemphasized issue in empirical prediction is that of stability. Especially when repeated application of the same method is envisaged, it is unlikely that the situations to be encountered will exactly mirror those involved in setting up the method. It may well be wise to use a procedure that works well over a range of conditions even if it is sub-optimal in the data used to set up the method.

1.11 Synthesis

It is reasonable to ask: what are the principles of applied statistics? The difficulty in giving a simple answer stems from the tremendous variety, both in subject-matter and level of statistical involvement, of applications of statistical ideas. Nevertheless, the following aspects of applied statistics are of wide importance:

- formulation and clarification of focused research questions of subject-matter importance;
- design of individual investigations and sequences of investigations that produce secure answers and open up new possibilities;
- production of effective and reliable measurement procedures;
- development of simple and, where appropriate, not-so-simple methods of analysis, with suitable software, that address the primary research questions, often through a skilful choice of statistical model, and give some assessment of uncertainty;
- effective presentation of conclusions; and
- structuring of analyses to facilitate their interpretation in subject-matter terms and their relationship to the knowledge base of the field.

All these aspects demand integration between subject-matter and statistical considerations. Somewhat in contrast, the role of work in the theory of statistics is to develop concepts and methods that will help in the tasks just listed. In pursuing this particular aim it is sensible to use sets of data to illustrate and compare methods of analysis rather than to illuminate the subject-matter. The latter use is entirely appropriate but is not the focus of discussion in the present book. Our emphasis is on the subject-matter not on the statistical techniques as such.

Notes

Detailed references for most of the material are given in later chapters. Broad introductory accounts of scientific research from respectively physical and biological perspectives are given by Wilson (1952) and by Beveridge (1950). For a brief introduction to the formal field called the philosophy of science, see Chalmers (1999). Any direct impact of explicitly philosophical aspects seems to be largely confined to the social sciences. Mayo (1996) emphasized the role of severe statistical tests in justifying scientific conclusions. For an introduction to data mining, see Hand *et al.* (2001). Box (1976) and Chatfield (1998) gave general discussions of the role of statistical methods.

2

Design of studies

This, the first of two chapters on design issues, describes the common features of, and distinctions between, observational and experimental investigations. The main types of observational study, cross-sectional, prospective and retrospective, are presented and simple features of experimental design outlined.

2.1 Introduction

In principle an investigation begins with the formulation of a research question or questions, or sometimes more specifically a research hypothesis. In practice, clarification of the issues to be addressed is likely to evolve during the design phase, especially when rather new or complex ideas are involved. Research questions may arise from a need to clarify and extend previous work in a field or to test theoretical predictions, or they may stem from a matter of public policy or other decision-making concern. In the latter type of application the primary feature tends to be to establish directly relevant conclusions, in as objective a way as possible. Does culling wildlife reduce disease incidence in farm animals? Does a particular medical procedure decrease the chance of heart disease? These are examples of precisely posed questions. In other contexts the objective may be primarily to gain understanding of the underlying processes. While the specific objectives of each individual study always need careful consideration, we aim to present ideas in as generally an applicable form as possible.

We describe a number of distinct types of study, each raising rather different needs for analysis and interpretation. These range from the sampling of a static population in order to determine its properties to a controlled experiment involving a complex mixture of conditions studied over time. In particular, the distinction between observational and experimental studies is central, in some ways more for interpretation than for analysis.

Despite these important distinctions, common objectives can be formulated. These are:

- to avoid systematic error, that is distortion in the conclusions arising from irrelevant sources that do not cancel out in the long run;
- to reduce the non-systematic, that is haphazard, error to a reasonable level by replication and by other techniques designed largely to eliminate or diminish the effect of some sources of error;
- to estimate realistically the likely uncertainty in the final conclusions; and
- to ensure that the scale of effort is appropriate, neither too limited to reach useful conclusions nor too extended and wasteful, as revealed by the achievement of unnecessarily high precision.

There are two further aspects, very important in some contexts, but whose applicability is a bit more limited. These are:

- by investigating several issues in one study considerably more information may be obtained than from comparable separate studies; and
- the range of validity of conclusions may be extended by appropriate design.

In the discussion in this book we concentrate largely on the careful design and analysis of individual studies aimed to reach an individually secure conclusion. Yet, while occasionally one comes across a unique largely stand-alone critical investigation, in most situations the synthesis of information from different kinds of investigation is required and this is an aspect that rarely lends itself to formal discussion. Even when conclusions are reached by the synthesis of results from related similar studies, as in so-called *overviews*, the quality of the individual studies may remain important.

In connection with overviews the term *meta-analysis* is sometimes used. The term may be misleading in the sense that no essentially different principles of analysis are involved, merely the difficult issue of deciding what data to include to ensure some comparability. More challenging still is a broad review of *all* aspects of a field, assembling information of different types as far as possible into a coherent whole.

Illustration: Adverse results end a line of enquiry   In a series of studies at
different sites, a modification of a textile process was found to produce
an improved product. Then in one further study disturbingly negative
results emerged. Although there were some questions over the design
of this last study it led to the abandonment of that whole line of investigation. Similar issues would arise if, in a series of studies of a new
medication, a therapeutic benefit appeared but one further study, possibly of questionable design, indicated a safety problem with the new
procedure. The point in both cases is that, while in a sense abandon-
ing the investigation because of one suspect study may be unreason-
able, it may be quite likely that such abandonment becomes almost
inevitable.

Especially in situations not motivated by an urgent decision-making requirement, investigation will proceed sequentially and much depends on
the time scales involved. If, as in some laboratory work, new investigations
can be set up and completed quite quickly then the design of individual
component studies is less critical, and surprising or ambiguous results can
be tested by immediate repetition. In other fields the soundness of each
component study is of much more concern and independent confirmation
is at best a long-term prospect.

A final important element of study design is the formulation of a plan of
analysis, especially for studies which are expensive and take a long time
to complete. Not only should it be established and documented that the
proposed data are capable of addressing the research questions of concern,
but also the main configurations of answers that are likely to arise should
be set out and the availability of the data necessary for interpreting such
patterns checked. The level of detail of analysis specified depends on the
context. To take a simplified example, it might be enough to specify that the
relationship between a response variable, y, and an explanatory variable, x,
will be studied without setting out a precise method of analysis. In others it
may be a quasi-legal requirement, for example, of a regulatory agency, to
specify both the precise method to be employed, for example, linear least-
squares regression, and also the level of confidence interval or significance
test to be used.

A simple example of the need to contemplate potential patterns of outcomes is that there may be a confident belief that outcomes in group A
will be higher than those in group B. The issues raised if the data point
in the opposite direction have already been mentioned in the illustration

'Explaining the unexpected' (p. 6). If new data can be obtained relatively speedily, this aspect of planning is of less concern.

Illustration: Explaining how badger culling increased cattle disease   A large randomized field trial was undertaken to estimate the effects of two different forms of badger culling on the incidence of bovine tuberculosis in cattle (Donnelly *et al.*, 2003, 2006; Bourne *et al.*, 2007). Badgers had long been regarded as a reservoir of infection and small studies suggested that culling badgers might be an effective way of reducing disease in cattle herds. One form of culling in the trial involved a series of localized culls, each following the identification of tuberculosis in a cattle herd. The other involved annual widespread culls.

Surprisingly, it was found that the localized culling approach led to increased disease incidence in cattle herds (Donnelly *et al.*, 2003). Detailed data had been collected on the locations of culled badgers and their infection status and data had been collected routinely on the incidence of disease in cattle herds. However, these data were not sufficient to explain the *increase* in disease incidence in cattle, and an additional study had to be undertaken to examine badger density and ranging behaviour in the areas subjected to culling and to matched areas not subjected to culling (Woodroffe *et al.*, 2006).

As noted in Section 1.4, even if pre-specified methods have to be used it is, however, crucial not to confine the analysis to such procedures, especially in major studies. There are two rather different reasons for this. First, careful analysis may show the initial method to be inappropriate. For example, in a simple application of linear regression, the transformation of variables may be desirable to deal with nonlinearity or with heterogeneity of the variance. More importantly and controversially, the data, or experience gained during collection of the data, may suggest new and deeper research questions or even, in extreme cases, abandonment of the original objectives and their replacement. The first reason, the technical inappropriateness of the original analysis, may not be particularly controversial. The second reason, a change in objectives, is more sensitive. In principle, conclusions that depend on a radical change in objectives in the light of the current data are particularly likely to need an independent confirmatory study.

The general point remains, however, that, while an initial plan of analysis is highly desirable, keeping at all cost to it alone may well be absurd.

While we are mainly addressing detailed issues of technique, successful design depends above all else on the formulation of important, well-defined, research questions and on the choice of appropriate material and the ability to obtain reliable data on the key features. The issues involved in the choice of measurements are discussed in detail in Chapter 4.

2.2 Unit of analysis

Many investigations have the broad form of collecting similar data repeatedly, for example on different individuals. In this connection the notion of a *unit of analysis* is often helpful in clarifying an approach to the detailed analysis.

Although this notion is more generally applicable, it is clearest in the context of randomized experiments. Here the unit of analysis is that smallest subdivision of the experimental material such that two distinct units *might* be randomized (randomly allocated) to different treatments. The primary analysis of such data will be based on comparison of the properties of units. Randomization is aimed partly at achieving correct interpretation of any systematic differences uncovered. The study of patterns of variation within units may be either of subsidiary interest, in establishing how best to assess inter-unit differences, or it may be of intrinsic interest, but in the latter case the conclusions do not have the protection of randomization and are subject to the additional possibility of biased interpretation.

Illustration: Unit of analysis; some examples   In a typical randomized clinical trial each patient is randomized to one of a number of regimens. Patients are thus the units of analysis. In a cross-over trial each patient receives one treatment for a period, say a month, and then after a gap a possibly different treatment for another month. The unit of analysis is then a patient-month. Similar designs are common in nutritional studies and in parts of experimental psychology, where the unit of analysis is a subject–period combination.

Illustration: Unit of analysis; large scale   In a community-based public health investigation, a study area is divided into communities, so far as is feasible isolated from one another. One of a number of possible health policies is then randomly allocated, so that all the families in a particular area receive, say, the same literature and possibilities for preventive care. Outcome data are then collected for each family. The primary unit of analysis is the community, with the implication that outcome measures should be based in some way on the aggregate properties of each community and a comparison of policies made by comparing communities randomized to different policies. This does not exclude a further analysis in which distinct families within the same community are compared, but the status of such comparisons is different and receives no support from the randomization.

In more complicated study designs there may be several units of analysis, as for example in split-plot experiments in agricultural field trials and comparable designs used in industrial experimentation on multi-stage processes.

In investigations that are not randomized experiments it is often helpful to consider the following question: what would the primary unit of analysis in the above sense have been, had randomization been feasible?

In general the unit of analysis may not be the same as the unit of interpretation, that is to say, the unit about which conclusions are to drawn. The most difficult situation is when the unit of analysis is an aggregate of several units of interpretation, leading to the possibility of *ecological bias*, that is, a systematic difference between, say, the impact of explanatory variables at different levels of aggregation.

Illustration: Unit of analysis: clash of objectives   For country- or region-based mortality data, countries or regions respectively may reasonably constitute the units of analysis with which to assess the relationship of the data to dietary and other features. Yet the objective is interpretation at an individual person level. The situation may be eased if supplementary data on explanatory variables are available at the individual level, because this may clarify the connection of between-unit and within-unit variation.

In the complementary situation where the unit of analysis is potentially smaller than the unit of interpretation then some details may be neglected at the former level.

In some applications the primary response is a curve against time showing, for example, the concentration in the blood of a radioactive marker following its injection. In some contexts the objective may be a detailed study of the typical shape of these curves and if possible their representation by a suitable nonlinear equation, perhaps derived from a differential equation representing the process involved. If, however, the objective is the comparison of responses in different groups of individuals, that is, units of analysis, treated in distinct ways then it may be more suitable to characterize the response curves by some simple measures, such as the peak response attained and the area under the response-time curve.

The general moral of this discussion is that it is important to identify the unit of analysis, which may be different in different parts of the analysis, and that, on the whole, limited detail is needed in examining the variation within the unit of analysis in question.

2.3 Types of study

An *observational* study is likely to comprise one or more of the following:

- secondary analysis of data collected for some other purpose;
- estimation of some feature of a defined study population which could in principle be found exactly, for example the number of animals of a specified species in a defined geographical area;
- tracking across time of such features; and
- determination of the relationship between various features of the study individuals, examined

  - at a single time point
  - across several time points for different individuals
  - across several time points for the same individual.

The studies listed above are observational in the sense that, although in some of them the investigators may have substantial control over what is measured, the system itself, in particular the potential explanatory variables, are not assignable in that way. For example, in a comparison of different treatment regimes for patients suffering from a particular disease, in an observational study the investigator would have no control over the allocation of a treatment to a particular patient.

By contrast in an *experimental* study the investigator would have essentially total control, in particular over the allocation of treatments. In the above example, some element of randomization would be typically be involved. See the illustration 'Conflicting observational and experimental evidence' (p. 7).

In some contexts an experimental approach, while in principle desirable, is impracticable or even unethical. In such situations a powerful start to the design of an observational study is to consider a hypothetical experiment and to design the observational study as far as possible to mitigate the differences between the observational and the experimental approaches, in particular to control any ambiguities of interpretation likely to arise in the former.

2.4 Avoidance of systematic error

We use the term *systematic error* rather than *bias* to avoid confusion with the much narrower term *unbiased estimate* as used in statistical theory. There are broadly two ways in which systematic error can arise. One is through the systematic aspects of, for example, a measuring process or the spatial or temporal arrangement of units and the other is by the entry of personal judgement into some aspect of the data collection process. The first source may often be avoided by design or by adjustment in analysis. The second source typically involves some element of randomization and very often an important element of concealment, called blinding.

Illustration: Benefit of a balanced design   Suppose that each run of an industrial pilot plant takes half a day. Two conditions are compared, a new treatment, T, and a control, C. All runs of T are in the morning and all runs of C in the afternoon; see Table 2.1a. However many replicate observations are taken, a comparison between the response variables for T and those for C is also a comparison of morning and afternoon conditions, and so there is the possibility of a systematic error. Unless, as is just conceivable, there were strong prior reasons for establishing the absence of a temporal effect, such a configuration would not be adopted in an experiment; typically the experiment would be arranged so that T and C both occur equally often in both positions; see Table 2.1c. In an observational study in which the undesirable extreme configuration arose, the only precaution against misinterpretation would be to look for external evidence that any systematic time effect is likely to be negligible; for example, some relevant measured features may show no systematic time effect. If, however, most but not all runs of T are in the morning then there is the possibility of eliminating a systematic effect by analysis. See Table 2.1b.

The analysis of Table 2.1b would typically be based on a model which, for approximately normally distributed quantitative outcomes, would best be written in a symmetrical form. Thus on day i the

Table 2.1 *Simplified form of comparison of T and C.*
(a) Extreme configuration; (b) unbalanced configuration;
(c) a standard experimental design

(a)

Day	1	2	3	4	5	6	7	8
morning	T	T	T	T	T	T	T	T
afternoon	C	C	C	C	C	C	C	C

(b)

Day	1	2	3	4	5	6	7	8
morning	T	T	T	C	T	T	C	T
afternoon	C	C	C	T	C	C	T	C

(c)

Day	1	2	3	4	5	6	7	8
morning	T	T	C	T	C	T	C	C
afternoon	C	C	T	C	T	C	T	T

observations y_{ij} are given by

$$y_{ij} = \mu + \tau a_{ij} + \delta b_j + \epsilon_{ij}, \tag{2.1}$$

where $j = 1$ for morning and $j = 2$ for afternoon; $a_{ij} = 1$ if T is used
and $a_{ij} = -1$ if C is used. Similarly, $b_1 = 1$ for morning observations
and $b_2 = -1$ for afternoon observations. The remaining terms ϵ_{ij} are
assumed to be independent random variables normally distributed with
mean zero and variance σ^2. There are many variants of such a model;
for example, other kinds of response variable may be used.

Suppose that in a study continuing over n days, where n is even, treatment T occurs pn times in the morning. Then application of the method
of least squares shows that, if $\hat{\tau}$ denotes the estimate of the treatment
parameter τ, which is one-half the difference between the means for T
and C, then

$$\mathrm{var}(\hat{\tau}) = \{8p(1 - p)n\}^{-1}\sigma^2, \tag{2.2}$$

where σ^2 denotes the variance of a single observation. Thus if $p \neq \frac{1}{2}$ then there is an increase in the variance of the estimated treatment effect following adjustment for a possible morning versus afternoon difference, although provided that the imbalance is not extreme the increase in variance is not great. For example, the appreciably unbalanced configuration in Table 2.1b would give $\mathrm{var}(\hat{\tau}) = \frac{1}{12}\sigma^2$, whereas the standard balanced configuration in Table 2.1c would give $\mathrm{var}(\hat{\tau}) = \frac{1}{16}\sigma^2$.

In more complex observational situations in which adjustment for several potential confounding features may be involved, the situation is more problematic.

The second type of systematic error arises when personal judgement enters, for example in a measuring process or indeed at any stage of an investigation. In many situations the use of randomization, that is, an impersonal allocation procedure with known probability properties, often combined with the so-called blinding of individuals, that is, concealment, is the most effective way of removing such sources of systematic error. This applies both in the sampling methods used in observational studies and in the design of experiments.

Illustration: Accurate assessment of precision   In an investigation of the reproducibility of a particular laboratory technique, duplicate samples are available from a number of somewhat different sources. If the duplicates are submitted for measurement very close together in time and especially if the result from the first measurement is known when the second measurement is taken, it is very likely that a misleadingly optimistic estimate of precision will result. The presentation of material not only in random order but with the source not being known to those performing the measurements is very desirable. Note that very often randomization without concealment would not be nearly so effective.

The need for randomization and concealment is even greater in measuring processes relying directly on personal judgement, for example in connection with taste testing. If randomization is not feasible then systematic allocation may be the best route to the avoidance of personal bias.

Illustration: Systematic sampling   To determine the twist in a textile yarn, a property difficult to measure, the yarn is examined at a series of sampling points. To space these randomly along the length of yarn under test would be cumbersome to implement. Instead the yarn may be advanced an exactly constant amount after each test, that is, the sampling points are exactly equally spaced along the yarn. The object of the enforced systematic aspect of the sampling is to avoid subjective biases in choosing the points for measurement. This assumes it to be known that a strictly periodic component of variation is unlikely.

In relatively complex situations, systematic error may enter at any of a number of phases and, for example, a single randomization may be insufficient.

2.5 Control and estimation of random error

Statistical analysis is particularly important in investigations in which haphazard variation plays an important role. This may enter at any stage of an investigation and a key aspect in analysis is the representation of that variability in a way that is as reasonably realistic and yet economical as possible. This will be a recurring theme of later chapters.

The steps in the design phase to lessen the impact of haphazard variation are essentially:

- use of artificially uniform material;
- arranging that the comparisons of main interest are, so far as feasible, of like with like;
- inclusion of background variables that will, in part at least, explain the haphazard variation encountered; and
- replication.

Illustration: Comparing like with like   The use of twins in a paired design for some kinds of animal experiment may greatly enhance precision by ensuring that conclusions are based largely on comparisons between the twins in a pair, according to the second principle for precision enhancement. The general disadvantage of using artificially uniform material is the possibility that the conclusions do not extend to more general situations.

Particularly in the initial phases of an investigation, it will usually be wise to study the phenomenon in question in situations that are as clear cut as is feasible. For example, in a study of possible control methods for an infectious disease it is sensible to recruit regions on the basis of high incidence, rather than with the aim of obtaining results representative of the whole population.

2.6 Scale of effort

In many investigations not only does the overall size of the investigation need consideration but also such details as the amount of replicate observation that may be desirable at various levels. This requires some knowledge of the relevant components of variance and their contribution to the overall uncertainty. All considerations of the scale of effort are in a sense economic, since the costs of observations must be balanced against the losses implicit in reaching conclusions of low precision. It is, however, probably rare that such considerations can be assessed fully quantitatively. Instead, a judgement is typically made on the basis of the level of precision likely to be achieved and, often, the level of precision thought fruitful in the field of work concerned. While judgements of scale of effort made this way involve arbitrary choices and are inevitably approximate, they do establish some comparability between different but related studies and so it is important that, especially for major investigations, some such calculations are made. In situations where resources for an investigation are limited, for example, the number of suitable patients for a clinical trial is small, the issue will be not so much calculating the size of study desirable as establishing whether the resources available and the number of patients likely to be accrued will be sufficient.

Such calculations are often presented in terms of the power of significance tests, which give the probability of detecting a preassigned departure from a null hypothesis at a specified level of statistical significance. In nearly all cases, however, it is simpler and probably more relevant to consider the standard error of the estimate of the primary aspect of interest in the study. If, for example, the latter is a comparison of two means each based on m independent observations, the value of m required to achieve a desired value c for the standard error of the estimated difference will be

$$\tilde{m} = 2\sigma^2/c^2, \tag{2.3}$$

where σ^2 is the variance of the observations in each of the groups being compared. To use this expression we need an approximate estimate of σ, obtained either from a pilot study or from experience in similar earlier investigations. Note that if the issue is formulated as requiring a power β at a specified distance d from the null hypothesis in a significance test at a one-sided level α then the corresponding value of m is

$$2\sigma^2 (k_\alpha^* + k_\beta^*)^2 / d^2. \tag{2.4}$$

Here k_α^*, for example, is the upper α point of the standard normal distribution.

There is one requirement, that of c, in the first formulation but in the second formulation three choices are required namely of α, β, and d, many of which will lead to the same value of m. A further reason for preferring the standard error is a general preference for estimation over significance testing and, importantly, that the standard error remains relevant for analysis whereas once the data are available the power calculation becomes largely irrelevant.

In most situations in which a number of qualitatively different treatments or exposures are under comparison, it is reasonable to aim for exact or approximately equal replication of the different treatments. An exception occurs when there is a control and a number, t, of other treatments, and interest focuses on comparisons of the other treatments, one at a time, with the control. It is then reasonable to have approximately $\sqrt{t}$ observations on the control for each observation on the other treatments. For example, with three new treatments this would require sets of five units, two for the control and one each for the other treatments. In fact even for two additional treatments over-replication of the control is often desirable.

2.7 Factorial principle

In some contexts, notably laboratory experiments that can be completed relatively quickly, it may be best to progress by a chain of simple experiments each informed by the preceding results. According to the *factorial principle*, however, in complex situations it may be desirable or even essential to examine several different aspects simultaneously. The issues involved are best illustrated by one of the first applications of factorial design.

Illustration: Factorial design   The growth of an agricultural crop is largely influenced by the availability of three components (factors): nitrogen, N; potassium, K; and phosphates, P. Consider an investigation in which each factor may be added to a plot at either a high or a low level and the resulting yield measured. Suppose that 24 plots are available. Two possible designs, each in blocks of eight plots, are as follows.

Table 2.2 *Three randomized replicates of the 2^3 factorial system. The notation 1 indicates that all factors are at their lower level, pk indicates the combination with N at its lower level and P, K at their upper levels, etc.*

Block 1	p	nk	1	k	npk	n	pk	np
Block 2	npk	n	nk	np	1	pk	p	k
Block 3	n	k	nk	pk	1	p	npk	np

- Eight plots are used to test N, eight to test P and eight to test K. In the plots used to test, say, N, the other two factors, P and K, are assigned at the low levels.
- The eight combinations of N, P and K at the two possible levels are each tested three times.

The second possibility, which may be described as three replicates of the 2^3 factorial system, is illustrated in randomized block form in Table 2.2.

There are two reasons for preferring the factorial arrangement in this context. One is that if, say, P and K have no effect (or more generally have a simple additive effect) then the estimated effect of N can be based on all 24 plots rather than merely eight plots and hence appreciably higher precision is achieved. More importantly in some ways, if there were to be a departure from additivity of effect there is some possibility of detecting this. An extreme instance would occur if any increase in yield requires all three factors to be at their higher level.

A development of this idea is that factors may also represent a classification of the experimental units rather than a treatment. For example, in the last illustration the experiment of Table 2.2 might be repeated, preferably but not necessarily with the same design independently randomized, in a number of farms in different regions with different soil characteristics. Replication of this sort would be essential if it were hoped that the conclusions would have an extended range of validity beyond the conditions of the single environment involved in Table 2.2.

We discuss factorial experiments further in Section 3.3.3. Similar ideas apply in observational studies. The investigation of a number of explanatory variables simultaneously may also have substantial advantages.

There is an extensive literature, discussed briefly in Section 3.3.4, on special designs for complex situations involving many factors. While these special designs have their uses, complex arrangements can have major disadvantages, in particular the difficulties of administering an intricate design and of slowness in obtaining complete results for analysis. Finally, we emphasize that the central principles of design apply as much to simple experiments as to complex ones.

Notes

Section 2.3. The terminology used for these different types of study varies between fields of work.

Section 2.4. R. A. Fisher (1926; 1935), who introduced formal randomization into experimental design, emphasized its role in justifying 'exact' tests of significance, that is, its specifically statistical role. In many contexts it can be used somewhat similarly to justify estimates of variance in both sampling and experimental design contexts. We have chosen here to emphasize the somewhat more qualitative and often more important matter of controlling conscious or unconscious selection effects.

Section 2.6. Instead of an estimated variance obtained by a pilot study it may be possible to use the properties of simple distributions, such as the Poisson, binomial and exponential distributions, in which the variance is a known function of the mean. For example, if totally randomly occurring point events are observed until n points have been observed then the fractional standard error of the mean rate, that is, the standard error divided by the mean, is $1/\sqrt{n}$. A rather rare exception, in which an economic calculation of the scale of effort was made, is the calculation by Yates (1952) of the amount of effort that it was reasonable to devote to fertilizer trials. The cost of the trials was compared with the loss consequent on a sub-optimal recommendation of the dressing of fertilizer.

3

Special types of study

This second chapter on design issues describes the main types of study in more detail. For sampling an explicit population of individuals the importance of a sampling frame is emphasized. Key principles of experimental design are discussed, including the factorial concept. Finally, various types of comparative observational investigation are outlined.

3.1 Preliminaries

We now discuss in a little more detail the main types of study listed in Section 2.3. The distinctions between them are important, notably the contrast between observational and experimental investigations. Nevertheless the broad objectives set out in the previous chapter are largely common to all types of study.

The simplest investigations involve the sampling of explicit populations, and we discuss these first. Such methods are widely used by government agencies to estimate population characteristics but the ideas apply much more generally. Thus, sampling techniques are often used within other types of work. For example the quality, rather than quantity, of crops in an agricultural field trial might be assessed partly by chemical analysis of small samples of material taken from each plot or even from a sub-set of plots.

By contrast, the techniques of experimental design are concentrated on achieving secure conclusions, sometimes in relatively complicated situations, but in contexts where the investigator has control over the main features of the system under study. We discuss these as our second theme in this chapter, partly because they provide a basis that should be emulated in observational studies.

3.2 Sampling a specific population

3.2.1 Sampling frame

Suppose that there is a specified target population and that one or more features of the population are of concern. In principle, even if not in practice, these features could be determined by measuring the whole population. In virtually all settings, however, it is necessary to estimate the population features by studying a subset, the *sample*. How should this sample be chosen?

Normally, the first requirement, of freedom from the possibility of systematic error, can be securely achieved only by eliminating individual choice by the investigator of the population members selected. This typically demands the existence of a *sampling frame*, that is, an explicit or implicit list of the population members.

A sampling frame of adult individuals in an area might be provided by the electoral register. If in an ecological study the population is a particular field or an area of forest, the map coordinates provide a sampling frame. For material passing along a conveyor belt at a constant speed, the times at which material passes a reference point provide a sampling frame. In so-called monetary-unit sampling for the auditing of accounts, the accounts are arranged in order and the cumulative sum of their values is formed. This produces a scale on which sampling takes place and so, implicitly, a sampling frame.

In the simplest forms of sampling, each member of the population has an equal chance of being chosen or, if there are unequal probabilities, these are known and adjustment is made in analysis. The reason for using probability-based methods is partly to ensure desirable statistical properties in the resulting estimates and, more specifically, to avoid systematic errors arising from personal judgment in the selection of individuals for inclusion. Systematic sampling, for example taking individuals at regular intervals in a list, may in many but not all cases largely avoid the latter.

Illustration: Monetary-unit sampling   Monetary-unit sampling, a widely used technique in the auditing of accounts, is illustrated in Figure 3.1. A starting point is taken at random, a sampling interval h chosen and those items falling within the marked points selected as the sample. Thus any item of value more than h is certain to be selected; otherwise the probability of selection is proportional to the value.

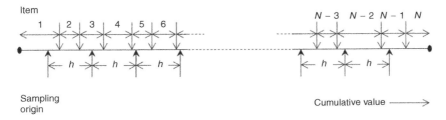

Figure 3.1 Monetary-unit sampling. The book values of items $1, \ldots, N$ are cumulated. The sampling interval is h. The sampling origin is chosen at random over $(0, h)$. A check is made of the items sampled in each interval h. Items with book values h and above are certain to be sampled. Otherwise the probability of inclusion of an item is proportional to its book value.

In a general formulation, if the ith individual in the population has a probability of selection π_i and the observed value relating to individual i is y_i then the mean value in the population is estimated by

$$\frac{\Sigma_s y_i / \pi_i}{\Sigma_s 1 / \pi_i}, \tag{3.1}$$

where Σ_s denotes summation over the sample. Indeed, for sampling discretely listed populations the specification of a sampling scheme sufficient for the estimation of means and variances depends upon the values of π_i and π_{ij}, where the latter is the probability that both individuals i and j are included in the sample. The probabilities π_{ij} directly determine the simpler probabilities π_i.

For monetary-unit sampling special issues are raised by the relevant objective, which is to study the incidence, typically small, of false values. In this context most or even all the recorded errors may be zero and this may call for special methods of analysis.

A general requirement is that either the probability of selection of a particular individual should not depend on the outcome variable for that individual or, if it does, that the dependence should be of known form and be allowed for in the analysis.

Illustration: Characterizing the distribution of lifetimes   To study the distribution of the lifetimes of a population of components in an engineering system, a sample is taken at a particular time and the lifetimes of the sample components determined. This sampling procedure distorts the

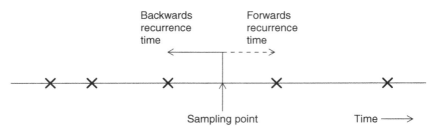

Figure 3.2 Recurrence times for a process of point events × occurring in time. The sampling points are chosen at random over long intervals. In the longer intervals, event are more likely to be captured. The recurrence times are measured from the sampling point backwards or forwards to the previous or succeeding events.

observed distribution because the chance of selecting a component of lifetime y is proportional to y. Thus the frequency function of the data is $yf(y)/\mu$, where $f(y)$ is the frequency function (probability density function) of the component lifetimes and μ is the mean of the distribution $f(y)$. This new distribution is formed by *length-biased sampling*. This is a special case of a sampling procedure in which the chance of selecting a particular individual depends in a known way on a property of that individual, which may happen by design or by force of circumstances.

Another possibility is that one measures the current age of the individual sampled, called the *backwards recurrence time*. There is a complementary definition of forwards recurrence time as the time elapsing before the next event in the process. The frequency distribution of this quantity is

$$\int_y^\infty f(z)\,dz/\mu. \tag{3.2}$$

See Figure 3.2. In all these cases the analysis must take account of the particular sampling method used.

3.2.2 Precision enhancement

There are two broad techniques for precision improvement: stratification and the use of control variables. In stratification the population is divided into distinct groups, strata, and the population properties of the strata, in particular the number of individuals in each stratum, are known. Then, in general by some combination of design and of analysis, the known distribution in the population is imposed on the sample. In some situations the strata are merely devices for guiding the sampling; in others there may be an interest in individual strata or in comparisons of different strata, in which case the strata are called *domains of study*.

Typical strata in studies of human populations are urban versus rural, male versus female and so on. In sampling material from an agricultural field trial for laboratory testing the domains of study might correspond to different treatments.

In general if there are k strata containing $(N_1, \ldots, N_k)$ individuals and data with means $(\bar{y}_1, \ldots, \bar{y}_k)$ from samples of sizes $(n_1, \ldots, n_k)$, the overall population mean is estimated by

$$\frac{\Sigma N_i \bar{y}_i}{\Sigma N_i} \tag{3.3}$$

with variance

$$\frac{\Sigma N_i^2 \sigma_i^2 / n_i}{(\Sigma N_i)^2}. \tag{3.4}$$

Here σ_i^2 is the variance in stratum i, provisionally assumed known although typically estimated, and it is assumed that the sample is only a small proportion of the population, so that the so-called finite population correction is negligible. It can be shown from these formulae that for a given total sampling cost the variance is minimized when n_i is proportional to

$$N_i \sigma_i / \sqrt{c_i}, \tag{3.5}$$

where c_i is the cost of sampling one unit in stratum i. Similar arguments may be used in more complicated situations. If the strata are domains of study then it may be desirable, for example, to sample the small strata more intensively than (3.5) suggests. In its simplest form, in which all strata have very similar statistical properties except for the mean, the optimal formula (3.5) forces an exact balance rather than the approximate balance induced by totally random selection. Even without optimal allocation, the analysis based on (3.3) and (3.4) corrects to some extent for any imbalance in the allocation achieved.

The other route to precision improvement is via the use of variables whose population distribution is known. Correction for any imbalance in the sample may then be made either by proportional adjustment or, often better, by a regression analysis. Let variable z have a known population mean $\bar{z}_\pi$ and a sample mean $\bar{z}_s$, and let variable y be assumed to be proportional to z. For example, if z were the number of individuals in a household, y might be a household's weekly expenditure on food. Then a proportional adjustment to the sample mean of the target quantity of interest, say $\bar{y}$, would change it to $\bar{y}\bar{z}_\pi/\bar{z}_s$. A regression adjustment would establish a regression relationship between the response variable y and the explanatory variable z and then use that relationship to estimate the mean of y at $z = \bar{z}_\pi$.

Adjustments to improve precision and remove potential systematic errors are common in the physical sciences. The adjustments may be based on well-established theory or be essentially empirical. Thus measurements of electrical resistance made at room temperature may be adjusted to a standard temperature using a known value for the temperature coefficient of resistivity of the material and the linear relationship between resistance and temperature known to hold over relatively narrow ranges. This is in effect a regression adjustment with a known rather than an estimated regression coefficient. Studies of the relationship between the tensile properties and the molecular structure of textile fibres are complicated by the fact that such fibres are not ideally elastic, so that their properties depend in particular on the rate at which stress is applied. Empirical correction to a standard rate may be needed, or results may be adjusted to a standard fracture time in order to produce comparability for different studies.

Although the details are different, these two methods of precision enhancement are broadly comparable with the use of randomized block designs and of precision improvement by covariates in the design of experiments.

3.2.3 Multi-stage and temporal sampling

These ideas can be extended in various ways when the population has a more complicated form. In multi-stage sampling the central idea of random sampling is applied at more than one level.

Illustration: Sampling frame If a population comprises all primary school children in a certain area, a sampling frame would be provided by a list of primary schools, lists of classes within these schools and then lists of children within classes. A sampling scheme would then specify a choice of schools within the area, a choice of classes within the selected schools and finally a choice of children within the selected classes.

In many adaptations of these ideas, individuals are followed over time. It may be wise in a complex study not to measure every feature at each sampling point; it may be helpful to use sampling with partial replacement of units and, at least in some situations, it may be necessary to pre-specify the sampling time points and the sampling intervals.

Illustration: Studying a panel of individuals In some studies, for example in studies of electoral behaviour, a panel of individuals is interviewed, typically at regular intervals, say yearly, in order to study changes over time in their potential voting preferences and attitudes. In sampling with partial replacement of units, at each new sampling point some individuals are omitted from the study and replaced by new individuals. Typically this has the dual object of reducing panel attrition by non-response and also increasing representativeness.

Illustration: Designing with multiple objectives In some contexts multiple objectives have to be addressed and this usually precludes the use of any sampling scheme that is optimal in a specifically defined sense. In a hydrological investigation 52 rain gauges were placed in a small river basin, which, for sampling purposes, had been divided into 2 km by 2 km squares; for a detailed discussion, see Moore *et al.* (2000). The objectives were to estimate the total hourly precipitation per square, essentially a problem of numerical integration, and to estimate local variability. In principle the latter could be achieved by comparing the values recorded at duplicate sampling points within a square, by comparing the values in different squares or by placing a large number of gauges in certain squares. The design adopted had 22 squares with one gauge, seven squares with two gauges and two squares each with eight gauges. While optimality calculations were used to a limited extent, especially in positioning the gauges in the two intensively sampled squares, a substantial arbitrary element was involved in balancing the differing objectives.

3.2.4 Less standard sampling methods

All the above methods presuppose a sampling frame. Without one the possibility of systematic misrepresentation of a population must be a serious concern, although analytical discussion of the relationship between the properties of study individuals may be less problematic.

In some studies of the genetics of rare human diseases, individuals having a particular disease, called *probands*, are identified by any method open to the investigator. Then close relatives of the proband are studied. In the analysis the data on the proband are not used. Even though the sampling procedure for finding a proband is ill-defined, so that the probands are not necessarily representative of a defined population, nevertheless it may be entirely reasonable to assume that interrelationships inferred from the relatives of the probands are stable and informative.

Snowball sampling, and more specifically respondent-driven sampling, uses similar ideas. In the latter, initial members of the population, say of injecting drug users, are found by any possible method. Data on each are recorded and then each member is given, say, three tokens to hand to others. If these enter the study then they in turn are given tokens, and so on. Some downweighting of the data from the first individuals may be used to lessen the dependence on the vagaries of the initial choice. The estimation of explicit population characteristics, in particular of the population size, requires strong and largely untestable assumptions.

In other situations, explicit and to some extent testable assumptions may aid the sampling procedure.

Illustration: Capture–recapture sampling   In capture–recapture sampling to estimate the size of a population of, say, animals in a defined area, m_0 animals are captured, marked and released. Later a new sample of size n_1 is captured and the number m_1 of marked individuals is noted. Under some strong assumptions the population size is estimated as $n_1 m_0 / m_1$. These assumptions are that not only is the population closed, that is, that birth, emigration, immigration and death are negligible, but all animals are equally likely to be selected at the second phase, in particular, independently of whether they were marked in the first phase. There are many extensions of the capture–recapture idea, although all involve some version of this last largely untestable assumption. One example where the impact of marking was tested is the toe clipping of frogs, which was found to reduce the probability of recapture significantly, raising doubts about the use of toe clipping for species that are of concern to conservationists (McCarthy and Parris, 2004).

If the sampling procedure is indefinite, as for example when it depends on voluntary enrolment, useful anecdotal information may be obtained but this is unlikely to be a good guide to population properties.

Illustration: Selective data collection   The UK Meteorological Office has compiled a list of severe and extreme storms of duration at least one hour occurring in the UK since 1886. This was produced from a combination of rain-gauge data and newspaper and other reports. The data collection is thus selective, in that events in remote parts of the country are relatively less likely to be covered and also that the extent of rain-gauge coverage has changed greatly over the period. The finding that the rate of occurrence of extreme events has been essentially constant over the period is thus hard to interpret.

In other such contexts where the objective is primarily to obtain data for comparative analysis, detailed representativeness of the target population may not be of critical importance. Nevertheless some comparison of population and sample properties may be helpful. For example, the placing of instruments for monitoring air pollution may be largely outside the investigators' control. For studying time trends and seasonal fluctuations the detailed placing may not be critical; for studying the relationship with health data the considerations might be different since matching with the distribution of the human population becomes more relevant.

3.3 Experiments

3.3.1 Primary formulation

The experiments to be considered here are comparative in the sense that their objective is to investigate the difference between the effects produced by different conditions or exposures, that is, *treatments*. The general formulation used for discussing the design of such experiments is as follows.

Experimental units and *treatments* are chosen. An experiment in its simplest form consists of assigning one treatment to each experimental unit and observing one or more *responses*. The objective is to estimate the differences between the treatments in their effect on the response.

The formal definition of an experimental unit is that it is the smallest subdivision of the experimental material such that any two different units may experience different treatments.

Illustration: Cluster randomization   In a clinical trial in which each patient is randomized to a treatment, the experimental unit is a patient. In a multi-centre trial in which all patients in the same clinic receive the same treatment, through a system known as cluster randomization, the experimental unit is a clinic. In the latter case the primary analysis compares the experience of whole clinics; the comparison of different patients within the same clinic may give useful supplementary information but raises more difficult issues of interpretation. See also the illustration 'Unit of analysis; large scale' (p. 18).

The first requirement of experimental design is the avoidance of systematic error and this is typically achieved by some form of randomization. In some situations, as in some physical or engineering contexts, in which haphazard variation is small and thought to be essentially completely random, a systematic arrangement may be employed instead, particularly in small experiments.

The pristine simplicity, in principle, of the interpretation of a randomized experiment hinges on the following argument. Consider the comparison of, say, two treatment regimes T and C for blood pressure control. Some individuals are randomized to T and the remainder to C, all other aspects remaining the same. A standardized outcome measure is recorded for each individual. If there is an appreciable difference between the outcomes in the two groups then either it is a consequence of the play of chance or it represents an effect produced by the distinction between T and C; there are no other possibilities. The former possibility can be assessed securely in the light of the known properties of the randomization procedure.

Illustration: Design in the presence of a trend   In a textile experiment on the effect of relative humidity in processing on the properties of yarn, nine runs were made. The runs occurred at weekly intervals and at one of three different controlled levels of relative humidity, 50%, 60% and 70%. In addition, because of the way in which the raw material was produced, it was thought possible that there would be a smooth, approximately quadratic, trend across the time period of the experiment. Randomization in such a situation may produce a very unsatisfactory configuration of treatments, and a systematic design allowing efficient least-squares estimation of the treatment effects and trend is preferable. This provides an instance where the design is formulated to achieve optimal estimation in a least-squares analysis. Randomization is confined to naming the treatments. In this context subjective biases can reasonably be neglected. Table 3.1 gives the specific design used.

Table 3.1 *Experimental design for three treatments in the presence of a temporal trend. There are nine equally spaced time points. The treatments, T, are the levels of relative humidity*

T_{60}	T_{50}	T_{70}	T_{70}	T_{60}	T_{50}	T_{50}	T_{70}	T_{60}

It can be shown that for this particular design a linear trend in time has no effect on treatment comparison and that a quadratic trend in time has no effect on the comparison of T_{70} with T_{50}; for other comparisons of treatments there is some mixing of effects, which may be resolved by a least-squares analysis.

Illustration: Protection against bias   In a randomized trial to assess the effect of culling badgers on the incidence of tuberculosis in cattle herds, randomization was used to allocate culling regimes to trial areas. The randomization was witnessed by an independent observer. The object was, at least in part, to protect the investigators against suggestions of conscious or unconscious bias in the study of what is in fact a contentious issue.

There are, however, potential complications. One common occurrence is noncompliance with the treatment allocated, particularly when this involves regular application over an extended period. Those allocated to T may not follow this regime and those allocated to C may indirectly follow a regime close to T. Analysis by *intention to treat*, sometimes called 'ever randomized, always analysed', ignores such complications and in effect compares *all* the consequences of being randomized to T with all the consequences of being randomized to C. It is clear that, at least in cases of extensive noncompliance, this can lead at best to an incomplete interpretation of the data. Satisfactory adjustment for noncompliance may require quite strong assumptions; it is in principle desirable that, if feasible, the reasons for noncompliance are recorded in each case. In some more general contexts, if major departures from the proposed design occur then it may be best to stop and restart the investigation afresh; when this is not possible it seems clear that the analysis should be of the experiment as actually performed!

Illustration: Post-randomization differences   New HIV prevention strate-
gies are being developed that include components such as the use of
vaginal microbicides. Clinical trials of these components include, as a
secondary intervention, condom provision and education to all randomized groups, to ensure that all trial participants are as far as possible
protected from HIV transmission. Complications arise in the interpretation of results from standard intention-to-treat analyses if condom use
differs between the randomized groups (Shiboski *et al.*, 2010). Adjustments to allow for post-randomization differences in condom use must
be based on strong, typically untestable, assumptions.

Randomization or, failing that, some other means of control, is needed at
every stage of an investigation at which appreciable error may enter. If any
element of personal choice by the investigator or other participant arises
then concealment becomes important, and this is often best achieved by
some element of randomization.

3.3.2 Precision improvement

In sampling an explicit population, precision is enhanced by the enforced balance of stratification or by the use of supplementary information. Broadly similar ideas apply in the design of experiments. The first, an
elaboration of the simple idea of comparing like with like, is to form blocks
of units expected on general grounds to have similar responses in the absence of treatment effects. The idea can be elaborated in various ways but
in its simplest version, the balanced randomized block design, the number
of units per block is the same as the number of treatments and the arrangement is randomized subject to the constraint that each treatment appears
just once in each block. The blocks may be formed in any way that the
investigator regards as convenient and likely to enhance precision by inducing balance. Note that personal judgement may and should be used to
the full in forming blocks. Randomization protects the investigation against
bias. Poor judgement in forming blocks will degrade precision but will not
itself induce bias.

A common way of forming a block is either by the amount of work that
can be done, say, in a day, by observers or sets of apparatus or by spatial
proximity.

The usefulness of such designs is not confined to continuous responses
analysed by the method of least squares, even though the analysis is

Table 3.2 *Comparison of degrees of freedom (DF) in a balanced randomized block design with b blocks, t units per block and t treatments. The total number of experimental units is bt*

Source	DF
mean	1
blocks	$b - 1$
treatments	$t - 1$
residual	$(b - 1)(t - 1)$

simplest in that case. We will discuss such an analysis for a design in b blocks, with t units per block and t treatments and thus $n = bt$ units in total.

The comparisons directly possible from this configuration are given in Table 3.2. Here the expressions in the second column indicate the numbers of logically independent contrasts possible on an additive scale. For example, among three things we may take the (independent) differences between A and B and between B and C; that between A and C is the sum of the separate differences.

This comparison structure suggests representation of the data through the sum of a general effect, a block effect, a treatment effect and a residual effect, to be considered temporarily as a source of error.

To be more specific still, if y_{is} is the observation in block i and is assigned to treatment s then it may be helpful to decompose the observations into the form

$$y_{is} = \bar{y}_{..} + (\bar{y}_{i.} - \bar{y}_{..}) + (\bar{y}_{.s} - \bar{y}_{..}) + (y_{is} - \bar{y}_{i.} - \bar{y}_{.s} + \bar{y}_{..}), \qquad (3.6)$$

where, for example, $\bar{y}_{.s} = \Sigma_i y_{is}/b$ is the mean observation on treatment s averaged over all blocks. It can be seen that the terms in the last set are all zero if and only if all observations are exactly the sum of a block term and a treatment term. The final set of terms is therefore called the *residual*. When set out in a $b \times t$ table, it will be seen that all rows and all columns sum exactly to zero by construction. Thus if $(b - 1)(t - 1)$ of the values are arbitrarily assigned, the table can be reconstructed and in that sense the residual is said to have $(b - 1)(t - 1)$ degrees of freedom.

Table 3.3 *Partitioned sums of squares (SS) in a*
balanced randomized block design with b blocks,
t units per block and t treatments. The total number of
experimental units is bt

Source	DF	SS
mean	1	$\Sigma_{i,s}\bar{y}_{..}^2$
blocks	$b-1$	$\Sigma_{i,s}(\bar{y}_{i.} - \bar{y}_{..})^2$
treatments	$t-1$	$\Sigma_{i,s}(\bar{y}_{.s} - \bar{y}_{..})^2$
residual	$(b-1)(t-1)$	$\Sigma_{i,s}(y_{is} - \bar{y}_{i.} - \bar{y}_{.s} + \bar{y}_{..})^2$

It follows, on squaring (3.6) and summing over all i and s, noting that all cross-product terms vanish, that we can supplement the decomposition of comparisons listed above as shown in Table 3.3.

Readers familiar with standard accounts of analysis of variance should note that the sums are over both suffices, so that, for example, $\Sigma_{i,s}\bar{y}_{..}^2 = bt\bar{y}_{..}^2$. At this point the sum of squares decomposition (Table 3.3) is a simple identity with a vague qualitative interpretation; it has no probabilistic content.

The simplest form of statistical analysis is based on the assumption that a random variable corresponding to y_{is} can be written in the symmetrical form

$$y_{is} = \mu + \beta_i + \tau_s + \epsilon_{is}. \tag{3.7}$$

Here the parameters β_i, typically of no intrinsic interest, represent the variation between different blocks of units whereas the τ_s correspond to different treatments; describing the contrasts between the τ_s is the objective. Finally, the ϵ_{is} are random variables in the simplest formulation independently normally distributed with constant and unknown variance σ^2. The numbers of block and treatment parameters can be reduced by one each to correspond to the number of independent contrasts available, but the symmetrical form is simpler for general discussion.

From this representation, application of the method of least squares shows that contrasts of treatments are estimated by the corresponding contrasts of treatment means, as expressed by comparisons among the τ_s, and that the residual variance σ^2 is estimated by the sum of squares of residuals divided by the residual degrees of freedom, $(b-1)(t-1)$. Confidence intervals for treatment effects are obtained by methods set out in any textbook of statistical methods. The sum of squares for treatments can be used to construct a test of the hypothesis that all treatments are equivalent; this is, however, rarely required.

We have given this simple description of the analysis of variance in order to stress the following points. First, and in some ways most importantly, the formalism of the analysis of variance sets out the comparisons that can be made from a particular type of design. This aspect does not depend on the type of variable measured or on specific assumptions about that data. Also, the general idea of an analysis of variance does not depend in an important way on the balance assumed in the randomized block design, that is, that each treatment appears once in each block. The analysis of variance table is often a valuable route to understanding and developing the analysis of relatively complex data structures. Next, there is the arithmetical decomposition of sums of squares. Finally, and it is only at this stage that a probability model is needed, there are statistical procedures for estimation and significance testing. These apply in the simple form given here for balanced data with continuous distributions and simple forms of error. It is important to note that, although the details become more complicated, analogous procedures apply to other sorts of model and to unbalanced data. An initial formulation of comparisons will show the broad form of model appropriate to the situation.

One extension of the randomized block principle arises when the units are cross-classified by two features, which we will call rows and columns. The most widely used arrangement of this kind is the Latin square, in which the experimental units are arranged in a $t \times t$ square and the treatment allocation is such that each treatment occurs once in each row and once in each column. Table 3.4 outlines the design of an investigation in experimental psychology in which eight subjects' reaction times are measured under four different stimuli, A–D. Each subject is tested in four sessions, suitably spaced in time, and the design ensures that the presence of systematic additive differences between subjects and between times do not degrade the precision of the treatment comparisons. The eight subjects are arranged in two separate 4×4 Latin squares; more squares could be added as appropriate.

The partition of information in the 32 observations is represented in Table 3.5. The new feature here, to be developed in more detail in the next section, is the difference in the roles of inter-subject variation and of inter-treatment and inter-period variation. The essential point is that the listing of subjects within each Latin square (section) is assumed to be arbitrary and possibly even randomized: the first subject in the first section has no

Table 3.4 *Design for eight subjects in two*
4 × 4 Latin squares, subject to stimuli A–D

Period	1	2	3	4
Square 1				
subject 1	B	C	A	D
subject 2	A	D	C	B
subject 3	D	A	B	C
subject 4	C	B	D	A
Square 2				
subject 5	D	A	C	B
subject 6	C	B	D	A
subject 7	B	D	A	C
subject 8	A	C	B	D

Table 3.5 *The degrees of freedom in a study*
of eight subjects in two 4 × 4 Latin squares
(for example, as in Table 3.4)

Source	DF
mean	1
Latin squares	1
subjects within Latin squares	3 + 3
periods	3
periods × Latin squares	3
treatments	3
treatments × Latin squares	3
residual within Latin squares	6 + 6

special connection with the first subject in the second section. Subjects are said to be *nested* within sections. However, the treatments, and indeed the periods, are meaningfully defined across sections; treatments are said to be *crossed* with sections. Thus the variation associated with subjects is combined into a between-subject within-section component. The treatment variation is shown as a treatment main effect, essentially the average effect across the two sections, and an interaction component, essentially showing the differences between corresponding effects in the two sections. More complex situations involving many factors are often best described by such a mixture of nesting and cross-classification.

Both in experiments and in roughly comparable observational studies, identifying the structure of the data in terms of treatments and explanatory features of the units and in terms of crossing and nesting is often the key to appropriate analysis.

The first route to precision improvement described above is by balancing, that is, comparing like with like. A second route is by measuring one or more explanatory variables on each unit, preferably before randomization, and, failing that, referring to the state of the unit before randomization. We may then adjust for imbalance in these variables either by performing stratification on them or by fitting a probability model in which the explanatory variable is represented through unknown parameters.

3.3.3 *Factorial experiments*

In Section 2.7 we illustrated a simple form of complete factorial experiment involving three factors N, P, K. We now illustrate some more general features. We consider a number of treatments A, B, ... each of which may be used at one of a number, $l_A, l_B, \ldots$ of possible levels. There are thus $l_A l_B \cdots$ possible treatment combinations and if each of these is used r times the experiment is called a complete $l_A \times l_B \times \cdots$ factorial experiment in r replicates. The illustration 'Factorial design' (p. 26) was thus three replicates of a $2 \times 2 \times 2$ or 2^3 factorial experiment.

Much discussion of factorial experiments centres on a decomposition into main effects and interactions of the contrasts possible between the different treatment combinations. This is best understood by repeated application of the ideas leading to (3.6). It leads in general to the decomposition of the treatment contrasts into a series of:

- main effects, involving $l_A - 1, \ldots$ degrees of freedom;
- two-factor interactions involving $(l_A - 1)(l_B - 1), \ldots$ degrees of freedom;
- three-factor interactions involving $(l_A - 1)(l_B - 1)(l_C - 1), \ldots$ degrees of freedom;
- and so on.

Thus the main effect of, say, treatment A, is a comparison of the responses at different levels of A averaged over all other factors. The two-factor interaction A × B indicates the departure from additivity of the means at different (A, B) combinations, averaged over the other factors, and so on.

Initial analysis of such balanced data will typically consist of a tabulation of one-way tables of means corresponding to the main effects, and two-way tables of means corresponding to the two-factor interactions. The discovery of appreciable interaction is typically a warning sign of departures from additivity of effect and these need careful study leading to specific interpretation; we return to this issue in Chapter 9.

3.3.4 Developments

There are many developments of the above ideas. They call for more specialized discussion than we can give here. Some of the main themes are:

- development of *fractional replication* as a way of studying many factors in a relatively small number of experimental runs;
- response-surface designs suitable when the factors have quantitative levels and interest focuses on the shape of the functional relationship between expected response and the factor levels;
- designs for use when the effect of a treatment applied to an individual in one period may carry over into the subsequent period, in which a different primary treatment is used;
- extensions of the randomized block principle for use when the number of units per block is smaller than the number of treatments (incomplete block designs);
- special designs for the study of nonlinear dependencies; and
- designs for sequential studies when the treatment allocation at one phase depends directly on outcomes in the immediate past in a pre-planned way.

3.4 Cross-sectional observational study

In some ways the simplest kind of observational investigation, one that is partly descriptive and partly analytical, is based on information in which each study individual is observed at just one time point, although at that same point information about that individual's past may be collected too. We will not discuss here the possible impact of recall bias, when the previous information is not collected from documented sources. We discuss issues of causal interpretation later, in Chapter 9, but note that, typically, to establish causality some notion of temporal flow is involved and therefore a causal interpretation of cross-sectional data is particularly hazardous.

Illustration: Interpreting a cross-sectional study   In a study of diabetic patients at the University of Mainz, described in more detail in Section 4.4 and by Cox and Wermuth (1996), data were collected for each patient on various psychometric measures, in particular the patient's knowledge of the disease. An objective measure of the success of glucose control was obtained at essentially the same time. In addition various demographic features and aspects of the patient's medical history were recorded. There was a clear association between knowledge and success at control. The data in this form being cross-sectional, it is not possible to conclude from this study on its own whether improvement in knowledge causes better control, whether success at control encourages better knowledge or whether some more complex process relates the two features. General background knowledge of medical sociology may point towards the former explanation, and its policy implication of providing more information to patients may be appealing, but direct interpretation of the data remains intrinsically ambiguous.

The data in the above illustration were obtained over an extended time period. In a rather different type of cross-sectional study a series of independent cross-sectional samples of a population is taken at, say, yearly intervals. The primary object here is to identify changes across time. More precise comparisons across time would be obtained by using the same study individuals at the different time points but that may be administratively difficult; also, if a subsidiary objective is to estimate overall population properties then independent sampling at the different time points may be preferable.

Illustration: Analysis of multiple cross-sectional studies   In a study of young people and obesity, data were analysed from three representative cross-sectional surveys (conducted in the UK in 1977, 1987 and 1997) (McCarthy *et al.*, 2003). The key variables under study were waist circumference and body mass index (BMI, calculated as weight in kilograms divided by the square of height in metres). Analysis showed that waist circumference increased dramatically over the period under study, particularly among girls. Smaller increases were observed in BMI but again increases were greater in girls than in boys. The authors were concerned, however, that the BMI gives no indication of the distribution of body fat, because central body fatness is associated with increased health risks.

3.5 Prospective observational study

We now consider observational studies in which the same individual is observed at more than one time point, leading to what is sometimes called longitudinal data. The term 'prospective observational study' is used when a group or cohort of individuals is followed forwards in time. There are a number of rather different possibilities. In some contexts the objective is to mirror a comparable randomized design as far as is feasible.

Illustration: Prospective observational study of long-term outcomes   Over many years, at Porton Down various biological agents were tested on volunteers from the UK armed forces. To study the possible effect on survival, and in particular on death from cancer, the volunteers formed the 'treated' group (Carpenter *et al.*, 2009). A comparable control group was formed from armed service members entering the forces near the same time point as assessed by their having adjacent record numbers, allocated on initial recruitment. This was not a randomized allocation but, subject to some checks, was reasonably assumed to be effectively random and so to lead to an estimated treatment effect free of systematic error. Therefore this study was reasonably close to a randomized experiment.

In structurally more complicated studies of this type the explanatory variables may themselves evolve in time.

In sociological studies of so-called event-history profiles, data are collected for each study individual of such events as completing full-time education, entering the labour market, obtaining employment, leaving employment, marriage, birth of children and so on. Here the objective is first to describe incisively the interrelationships between the various transitions of state involved and then, if possible, to detect systematic patterns of dependence. The design problems concern the choice of baseline data, such as the demographic details, to be obtained for each individual. For recording critical events it is necessary also to determine the sampling interval.

3.6 Retrospective observational study

A retrospective investigation is particularly relevant when a prospective study, while in principle desirable, is likely to be very inefficient because few individuals experience the critical event which is the outcome of interest. In the type of prospective study of concern here, we start with a group of individuals of interest, record appropriate explanatory features and wait for the response of interest to occur; in an epidemiological context this might be death or the occurrence of a heart attack. There are two connected disadvantages of such an approach. The first is that one may have to wait a considerable time to accrue an appreciable number of, say, deaths. The second is that at the end we will have collected an unnecessarily large amount of information on the non-cases.

By contrast, in a retrospective study we start with cases and for each case choose one or more controls. Either these may be chosen at random from the relevant population or each may be matched to a case with respect to features causally prior to the explanatory variables of concern. Thus the explanatory features are determined retrospectively. This leads to a much more balanced and efficient distribution of effort in data collection but is subject to the often serious disadvantage of the possibility of recall bias.

Put succinctly, a prospective study looks for the effects of causes whereas a retrospective study examines the causes of effects.

The statistical argument that underpins a case-control study is, in its simplest form, as follows. Consider a population with one binary explanatory variable, z, taking values 0 and 1 and a binary outcome, y, also taking values 0 and 1. In the population there are thus probabilities π_{is} corresponding to $z = i$, $y = s$. See Table 3.6(a). In a cohort study separate samples are drawn from the subpopulations $z = 0$ and $z = 1$ and the outcome is observed; this leads to the conditional probabilities shown in Table 3.6(b). By contrast, in a case-control study separate samples are drawn from the subpopulations with $y = 0$ and $y = 1$ and the corresponding values of z are observed; this leads to the conditional probabilities in Table 3.6(c).

The probabilities specified in Tables 3.6(b) and (c) are obtained by conditioning the probabilities in the population study in Table 3.6(a) to represent respectively sampling based on z and on y.

The link between Tables 3.6(b) and (c) and therefore between cohort and case-control studies is that the cross-product ratios for both tables, $(\pi_{11}\pi_{00})/(\pi_{01}\pi_{10})$, are the same as that for Table 3.6(a). This implies that assessment of the effect of z on y as obtained from the cohort study can also be estimated from the case-control study by treating y as an outcome variable, even though in this sampling process it was controlled. The conclusion is restricted to this particular measure of comparison. In a more general context it depends on the use of logistic regression.

Case-control studies are widely used in epidemiology. In econometrics they were used initially in a study of modes of transport under the name *choice-based sampling* (Ben-Akiva and Leman, 1985). In this particular

Table 3.6 *Distribution of a binary explanatory variable, z, and a response variable, y, in (a) a population study, (b) a prospective or cohort study, (c) a retrospective or case-control study*

(a) Population study

	$y = 0$	$y = 1$
$z = 0$	π_{00}	π_{01}
$z = 1$	π_{10}	π_{11}

(b) Prospective study

	$y = 0$	$y = 1$
$z = 0$	$\pi_{00}/(\pi_{00} + \pi_{01})$	$\pi_{01}/(\pi_{00} + \pi_{01})$
$z = 1$	$\pi_{10}/(\pi_{10} + \pi_{11})$	$\pi_{11}/(\pi_{10} + \pi_{11})$

(c) Retrospective study

	$y = 0$	$y = 1$
$z = 0$	$\pi_{00}/(\pi_{00} + \pi_{10})$	$\pi_{01}/(\pi_{01} + \pi_{11})$
$z = 1$	$\pi_{10}/(\pi_{00} + \pi_{10})$	$\pi_{11}/(\pi_{01} + \pi_{11})$

US investigation there were few cyclists, who were therefore heavily sampled.

Illustration: Retrospective case-control study In a long series of international investigations of the possible impact of radon gas in the home on lung cancer, the cases were hospital patients with lung cancer (Darby *et al.*, 2005). For each such patient their places of residence over the preceding 20 years were identified and either radon-measuring counters placed in these homes for a suitable period or the likely intensity of radiation in them was interpolated from data collected in nearby residences. In such investigations, comparable data are needed on controls and there are various ways in which these may be obtained, for example from hospital data on patients with an unrelated disease, from general practitioner (primary care physician) records or by sampling of the whole population using an appropriate sampling frame. Also, the controls may be individually matched to cases, say, by age and sex. All approaches have their advantages and disadvantages but we will not discuss them here.

Notes

Section 3.1. There are considerable, and largely disjoint, literatures on sampling and on the design of experiments. The separateness of these literatures indicates the specialization of fields of application within modern statistical work; it is noteworthy that two pioneers, F. Yates and W. G. Cochran, had strong interests in both topics. There are smaller literatures on the design of observational analytical studies.

Section 3.2. A general introduction to sampling problems was given by S. K. Thompson (2002); a more theoretical discussion was given by M. E. Thompson (1997). Examples of industrial sampling, including an account of the statistical aspects of length-biased sampling, were given by Cox (1969) and of monetary-unit sampling in auditing by Cox and Snell (1979). An account of stereology, emphasizing the connection with traditional sampling problems, was given by Baddeley and Jensen (2005). For sampling in human genetics, see Thompson (2002) and for respondent-driven sampling see Gile and Handcock (2010).

Section 3.3. Modern discussions of experimental design stem from Fisher (1926; 1935). A non-technical account strongly emphasizing design rather than analysis is due to Cox (1958). An excellent modern account of these ideas is that by Bailey (2008). For a wide-ranging, more mathematical account, see Cox and Reid (2000). The response-surface aspects were emphasized by Box, Hunter and Hunter (1978). For an account of the very elaborate textile experiment within which the trend-balanced design of Table 3.1 was imbedded, see Cox (1952). The analysis of variance has various meanings, of which the most literal is the decomposition of random variations into component sources. The meaning discussed here of the decomposition of contrasts in complex experimental arrangements may seem a moribund art form. This aspect is, however, concerned with the structure of data, not merely with deduction from a contrived linear model, and is certainly not primarily a route for significance tests; hence it is of broad concern. This approach can best be followed from one of the older books on statistical methods, for example Snedecor and Cochran (1956). A more mathematical account was given by Scheffé (1959).

Section 3.5. For a systematic account of prospective studies in an epidemiological context, see Breslow and Day (1987).

Section 3.6. For a systematic account of case-control studies, see Breslow and Day (1980) and for an econometric perspective, see Amemiya (1985).

4

Principles of measurement

The success of an investigation depends crucially on the quality and relevance of the data analysed. Criteria for good measurement procedures are outlined and the quantities measured are classified in various ways, in particular by their role in the study in question.

4.1 Criteria for measurements

A crucial aspect of any analysis is the nature and quality of the data that are involved, which we call *measurements*, using that term in a very general sense. Criteria for satisfactory measurements typically include:

- relevance;
- adequate precision;
- economy; and
- absence of distortion of the features studied.

In fields with a solid history of investigation, the measurement techniques necessary to capture the relevant features may well be firmly established. In other contexts, defining a measurement procedure that will yield insightful information may be crucial to successful work. Usually we require of a procedure some mixture of face validity, that is, apparent relevance, and construct validity. The latter implies that a procedure has proven success in establishing meaningful and stable relationships with other variables. These further variables should include those for which there is strong prior reason to expect a connection.

The costs, however measured, of obtaining the data should be commensurate with the objectives. In particular, while individual data quality is of great importance, the collection of unnecessarily large amounts of data is to be avoided; in any case the intrinsic quality of data, for example the response rates of surveys, may be degraded if too much is collected. The use of sampling may give higher quality than the study of a complete population of individuals.

Illustration: Questionnaire data   When researchers studied the effect of the expected length (10, 20 or 30 minutes) of a web-based questionnaire, they found that fewer potential respondents started and completed questionnaires expected to take longer (Galesic and Bosnjak, 2009). Furthermore, questions that appeared later in the questionnaire were given shorter and more uniform answers than questions that appeared near the start of the questionnaire.

However, human behaviour is complicated. Champion and Sear (1969) found that their longer questionnaires were returned significantly more often (nine-page questionnaire: 39%; six-page: 38%; three-page: 28%), even though the content was the same in all cases. Only the spacing between questions varied. A possible explanation is that the three-page questionnaire appeared unappealingly cluttered.

Finally, special precautions may be needed to minimize any direct or indirect effect of the measurement process on the system under study.

Illustration: Measurement affects the system studied   Studies of animal behaviour may require considerable efforts to minimize the presence of an observer on the effect under study. For example, a group of wild, non-habituated, common marmosets (*Callithrix jacchus*) were observed in 12 sessions (de Almeida *et al.*, 2006). In half the sessions two observers, dressed in camouflage clothing, maintained visual contact with the animals from 8 meters away, while in the other half the two observers were hidden inside a blind. The mean frequency of marmoset alarm calls was more than 10 times greater in the sessions where the observers were visible, with a mean of 8.65 calls per 10 minutes, than when they were observing from within the blind, when the mean was 0.77 calls per 10 minutes.

Indeed, this effect is not limited to studies of wild animals. The presence of observers has been found to affect a wide range of aspects of human behaviour including eating (Herman *et al.*, 1979), responding to pain (Sullivan *et al.*, 2004), teachers' classroom behaviour (Samph, 1976) and even which toys children choose to play with (Wilansky-Traynor and Lobel, 2008).

Another aspect is that the adoption of a particular measure of system performance as a performance target may distort the whole behaviour (Bird *et al.*, 2005).

Illustration: Performance targets distort   Shahian *et al.* (2001) described various forms of 'gaming' in hospital management, including the *up-coding* of mortality risk factors (comorbidities) to increase the predicted mortality rate for patients and thus increasing the chance that the observed mortality rate for patients will be less than or equal to that expected. They also described instances in which high-risk patients were more likely to have their operations recorded as procedures for which mortality rates were not published. Finally, they described a tendency to transfer the patients most critically ill after surgery to other facilities in anticipation of their impending deaths when databases included only deaths occurring in the setting in which the surgery took place.

Illustration: Hawthorne effect   An early study of industrial management (Roethlisberger and Dickson, 1939) found what is now called the Hawthorne effect. In its simplest form this showed that a modification of a work process produced an improvement in productivity and a further modification a further improvement; a final modification back to the initial state produced yet more gain. The general moral to be drawn is that it is the act of intervention rather than the nature of the intervention itself that produces change, although this interpretation is controversial.

4.2 Classification of measurements

Variables may be classified in various ways, in particular by:

- mathematical properties of the measurement scale;
- whether they have been censored or otherwise degraded;
- their purpose in the investigation in question;
- whether they are direct or so-called derived measures; and
- whether they are observed in the system under study or are latent, that is, either they are unobserved because they are purely abstract, like ability at arithmetic, or they might have been observed but in fact were not.

The primary concern regarding the quality of measurements is to ensure that the study under discussion reaches secure conclusions. Often, a second and important consideration, however, is to achieve comparability with related investigations, in particular to connect with the research literature in the field. This can be a cause of tension, in that sometimes the standard methods of measurement, and indeed of analysis, in a field are ripe for worthwhile improvement.

Perhaps particularly in the physical sciences, the availability of internationally standardized measuring techniques of high precision has often underpinned spectacular progress. At another extreme, skilful special choice and definition may be involved. Thus in statistical literary studies of, for example, disputed authorship, counts of features judged to be context free are required. By contrast, *content analysis* emphasizes meaning and aims to interpret the intrinsic subject-matter of complex material into associated frequency distributions which can then be used for comparative purposes.

4.3 Scale properties

The most immediate classification of measurement procedures is by the nature of the set of possible values, according to whether they are:

- binary, that is taking two possible values, for example success and failure;
- qualitative, with no implied ordering, for example religious affiliation;
- ordinal, for example socio-economic status (high, middle or low);
- integer-valued, typically counts of occurrences;
- continuous, with or without a meaningful scale origin; or
- functional responses.

By a functional response is meant a measurement of a function, particularly a function of time. This differs from the ecological usage, in which the relationship between the density of a predator and the density of its prey is called the functional response.

Illustration: Growth curves   Growth curves of animals or plants may be summarized for further analysis by:

- fitting a parametric response versus time equation and characterizing each response curve by the estimates of key parameters;
- fitting a smooth function not specified parametrically; or
- characterizing each curve by descriptive features such as the rate of growth over an initial period, the apparent asymptote of the curve and so on.

These distinctions affect primarily the techniques of detailed analysis to be used rather than the key questions to be investigated. Moreover, while the distinctions are fairly clear cut, it may be wise to interpret them flexibly. For example, on scales that are in the strict sense ordinal the borderlines between the different possible outcomes are rarely totally arbitrary, so that an assumption that such scale values as are found in questionnaire answers, for example, very bad, bad, neutral, good, very good, can be treated as numerical, say, -2, -1, 0, 1, 2, may be a reasonable basis for at least a tentative analysis.

In some contexts a particularly important distinction is between extensive and intensive variables. Extensive variables (or properties) have the property of physical additivity, usually in virtue of their depending on the size of the system or sample studied. Examples include mass, volume and heat (thermal energy). By contrast intensive variables have no such additive feature. Examples include temperature and density. A particular sort of intensive property is a *specific property*, which is a property defined per unit mass. For example the specific volume, which is the volume occupied per unit mass (e.g. cubic meters per kilogram), is a specific property.

Illustration: Extensive variables   A standard example from physics is that quantity of heat is extensive and temperature is not. If two bodies with differing quantities of heat (thermal energy) and temperatures are allowed to come to thermal equilibrium with one another, the quantity of heat adds but the temperature neither adds nor averages. The yield of a product and the quantity of money are additive as are, at least in some contexts, length and duration.

An implication is that whatever the form of the distributions involved the arithmetic mean on the extensive scale has a specific interpretation.

Illustration: Interpreting an extensive variable   Often a set of non-negative continuous variables may best be analysed on a log scale, especially if the variables have approximately log normal distributions. But if an outcome variable is extensive, at least some aspects of the interpretation will be needed in terms of the mean on the original scale. For instance, in some contexts individual earnings might be treated as approximately log normally distributed and some interpretations made in terms of log earnings. Yet if the requirement were to consider family or aggregate community earnings, it would be the mean earnings not the mean log earnings that would be relevant.

Illustration: Ratio and difference of two probabilities   In epidemiology the effect of exposure on the occurrence of a disease is often measured by the relative risk, that is, the ratio of the probability of the disease in exposed individuals and the probability of the disease in otherwise similar unexposed individuals. That ratio, or in practice its logarithm, may well show statistically stable relationships with other explanatory features. Yet in assessing the effect of, say, eliminating exposure in a specific population it is the *difference*, not the ratio, of two probabilities that would determine the numbers of individuals affected by the change. The probability in this sense being extensive, it is differences in probabilities that have an immediate interpretation. Ratios often have other important advantages, such as stability across different populations and as a guide to potentially causal interpretation, so that analyses on both scales may be required.

4.4 Classification by purpose

In many ways a particularly important classification of measurements is by the role or roles of the variables in the study under analysis. A central idea is that, in any given context, for each pair of variables either one can be regarded as a response to the other as explanatory or the two variables are to be treated on an equal standing. This, with some additional restrictions, leads to the sorting of variables into groups in which all variables in a given group are on an equal standing. Furthermore the groups can be placed in an order preserving the direction of dependence, so that all the variables in one group are potentially explanatory to all the variables in another group; see Figure 4.1.

This leads to a broad classification of variables as response variables (or outcomes) or as explanatory variables.

Illustration: Response and explanatory variables   In the study of diabetic patients described in the illustration 'Interpreting a cross-sectional study' (p. 47) and by Cox and Wermuth (1996), the final response variable is success at controlling the disease as measured by a biochemical test. The primary explanatory variable is knowledge of the disease, assessed by a standard psychometric test. Explanatory to this variable are measures of disease attribution, assumed to influence motivation to attain knowledge. Finally, variables such as age, gender, years of

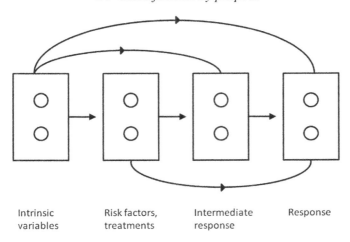

| Intrinsic variables | Risk factors, treatments | Intermediate response | Response |

Figure 4.1 Simple regression chain. Circles represent variables. Variables in the same box are of equal standing, i.e. none is a response to or explanatory to another. Variables in a given box depend in principle on all variables in boxes to their left. The effect of treatment on response is obtained conditionally on the variables to the left and ignoring the variables intermediate between treatment and response. The total effect may be decomposed, and possibly explained, according to the pathways between treatment and response.

education and years for which the patient has had the disease in effect define the individuals concerned. An interesting feature of this example is that the biochemical test and the psychometric measurements were obtained virtually simultaneously, so that the labelling of the biochemical measure as a response to knowledge is a subject-matter working hypothesis and may be false. That is, success at controlling the disease might be an encouragement to learn more about it.

The example illustrates some further distinctions. Some explanatory variables, such as gender or education, characterize the study individuals and are to be regarded as given. We call them *intrinsic*. In more formal analyses they are typically not represented by random effects but regarded as fixed characteristics.

Other explanatory variables might in a randomized experiment be treatments or conditions under the investigator's control, and the objective is to study the effect of these on the final outcome or response. Such variables are called *primary* explanatory variables. Conceptually, for each study individual a primary explanatory variable might have been different from the choice actually realized.

A third kind of explanatory variable, a particular type of intrinsic variable, may be called *nonspecific*. Examples are the names of countries in an international study or the names of schools in an educational study involving a number of schools. The reason for the term is that, depending of course on context, a clear difference in some outcome between two countries might well have many different explanations, demographic, climate-related, health-policy-related etc.

Other variables are *intermediate* in the sense that they are responses to the primary explanatory variables and explanatory to the response variables.

At different stages of analysis the roles of different variables may change as different research questions are considered or as understanding evolves.

Illustration: Classification of variables In an agricultural fertilizer trial the fertilizer applied to an individual plot is the primary explanatory variable, the number of plants per m^2 is an intermediate variable and the yield of the product is the final outcome or response.

A surrogate endpoint is a particular sort of intermediate variable that is used, for example, in clinical trials to compare the impact of treatments on the final outcome, or true endpoint. Prentice (1989) defined a surrogate endpoint to be 'a response variable for which a test of the null hypothesis of no relationship to the treatment groups under comparison is also a valid test of the corresponding hypothesis based on the true endpoint'. In this way the designation of a variable as a surrogate endpoint depends on the treatments under study. In practice, variables may fail as surrogate endpoints for a number of reasons, including the following (Fleming and DeMets, 1996):

- the variable might not be part of the same process as the true endpoint;
- there might be multiple pathways between the disease under study and the true endpoint, so that the treatment only affects pathways not involving the proposed surrogate endpoint or it only affects a pathway that does involve the proposed surrogate endpoint; and
- the treatment might affect the true endpoint independently of the disease process.

Illustration: Surrogate endpoint   In the study of drugs for the treatment of glaucoma, the intraocular pressure has been found to be a useful surrogate endpoint for the true endpoint of interest, vision loss. Treatments which lower intraocular pressure have been consistently found to reduce the risk of the loss of visual field (Vogel *et al.*, 1990).

4.5 Censoring

When failure time data (or survival data) are collected, often some units of analysis do not have fully observed time-to-fail data. Other kinds of data too may have measurements constrained to lie in a particular range by the nature of the measuring process. The most common occurrence is *right censoring*, in which at the end of the period of data collection some units of analysis have not yet failed and thus, at least theoretically, would fail at some future point in time if observation were continued.

The next most common type of censoring is *interval censoring*, in which the exact failure time is not observed but is known to have occurred within a particular interval of time. Such data are called interval-censored or *doubly censored*.

Finally, *left censoring* occurs when, for some units of observation, the time to fail is known only to have occurred before a particular time (or age).

Survival analysis is very widely used for right-censored data on the time to death, the time to disease onset (such as in AIDS diagnosis) or the duration of a health event such as a coma. Interval-censored data frequently arise in clinical trials in which the outcome of interest is assessed only at particular observation times; such data might be for example a biomarker dropping below a particular value or the appearance of a medical feature not readily recognized by the patient which might be assessed at a visit to the healthcare provider.

Illustration: Right-censoring   Investigations of the strength of materials and of their structures are commonly made by exposing them to repeated cycles of intensive stress. The resulting behaviour may be of intrinsic interest or may be a surrogate for behaviour under more normal conditions, an often reasonable assumption if the failure mode remains determined by the same physical or biological process. In this latter context it may be reasonable to stop testing, that is, to right-censor, all individuals that have survived a stress equivalent to the maximum likely to arise in applications.

However, censoring is not limited to survival data. Studies of HIV/AIDS often involve the analysis of quantitative plasma HIV-viral-load measurements. The datasets are often complicated by left censoring when the virus is not detectable, that is, the viral load is somewhere between 0 and the lower limit of quantification for the test used.

Similar challenges arise in the study of environmental contaminants when minimum detection levels in assays give rise to left-censored data (Toscas, 2010; Rathbun, 2006).

In economic datasets, data on wages may be right-censored (or *topcoded*) because the data are derived from a system, such as a social security system, with an upper limit (Büttner and Rässler, 2008). Thus, the highest wages are known only as being at least a particular value.

The statistical analysis of censoring is virtually always based on the often untestable assumption that censoring is in a special sense uninformative. That is, given the current observations at the time of censoring, the choice to censor the individual is conditionally independent of the unobserved potential failure time.

4.6 Derived variables

A further distinction is between variables that are directly measured, sometimes called *pointer readings*, and those formed from combinations of such variables. There are a number of types of derived variable.

One type is a combination of a variable with another variable or variables explanatory to the first. The object is to achieve some simplification and standardization.

Illustration: A derived variable   Studies of individual obesity often combine weight and height into the body mass index (BMI), the weight in kilograms divided by the square of the height in metres. Here height is regarded as in part explanatory to weight, and, except in infants, as an intrinsic variable, and the BMI is an attempt to achieve standardization in comparisons of different groups of individuals. That is, in a study in which body mass is a primary outcome, use of the BMI is a possibly crude attempt to standardize by height. A better procedure is likely to be to treat the log of the weight as a response and to regress on the log of the height as an intrinsic variable; a regression coefficient close to -2 indicates virtual equivalence to the use of the BMI.

By contrast, if all variables are on an equal standing then the derived variable is both a simplification and a route to a more sensitive conclusion, possibly obtained by combining measurements of limited individual accuracy.

Illustration: Principal components analysis   Principal components analysis is a particular mathematical technique for obtaining derived variables in the form of linear combinations of the original directly measured variables. These derived variables are uncorrelated and are typically reported in order of importance (the most important being the one that explains the most variation in the directly observed data). Nearly 2000 children (aged 6 to 10 years) participated in a study of children's fears that identified specific fears and their intensity (Salcuni *et al.*, 2009). Four factors were identified using principal components analysis:

1. death and danger;
2. injury and animal;
3. failure and criticism; and
4. fear of the unknown.

Girls were found to be significantly more fearful than boys.

In general, while principal component analysis may be helpful in suggesting a base for interpretation and the formation of derived variables there is usually considerable arbitrariness involved in its use. This stems from the need to standardize the variables to comparable scales, typically by the use of correlation coefficients. This means that a variable that happens to have atypically small variability in the data will have a misleadingly depressed weight in the principal components.

In some time series collected by official statisticians, data are obtained from sample surveys both to monitor issues of public policy and to form a base for social science research into such matters as crime, unemployment and so on. These series may continue over a substantial time period. Definitions, even of such basic variables as social class, may therefore reasonably change over time. In principle, when such changes occur both the new and old versions should be recorded over a transition period. Unfortunately this overlap may not take place; then special precautions may be needed in analysis.

> **Illustration: Price indices** The UK Office for National Statistics calculates the Consumer Prices Index (CPI) and the related Retail Prices Index (RPI) to reflect the state of the UK economy (Office for National Statistics, 2010). These variables reflect the price of a basket of goods on a particular date. The contents of the basket are reassessed each year, and the prices of the goods it contains vary, as estimated from price data collected throughout the UK. The weights given to the goods in the basket (and their prices) are chosen to reflect the spending of a typical household. The change, implemented in 2010, from RPI to CPI in tax and benefit calculations was particularly important since it had negative implications for many individuals in such respects as tax allowances, state benefits and pensions.

4.7 Latent variables

4.7.1 Generalities

The formulations discussed in earlier parts of this chapter concerned directly observed features, that is to say, these formulations are representations of data that have been or might be collected. In some contexts, however, it may be fruitful, or even essential, to consider *latent features*, that is, aspects that have not been, or sometimes even in principle cannot be, observed. The general ethos of much statistical thinking is to stay as close as is feasible to empirical data and therefore to avoid latent variables as far as possible, but there are a number of situations where use of the latter is almost unavoidable.

Thus the empirical study of classical dynamics is not based solely or even primarily on direct measurements of distance, time and mass but involves such latent variables as the potential and kinetic energy and the momentum. Other areas of physics abound with such concepts, some quite esoteric. While direct emulation of physical concepts, for example analogies between social geographical issues and the notion of gravitational potential, may be unduly forced, some use of latent features in theory construction seems unavoidable.

Latent variables are those which cannot be measured directly. Variables which can be so measured are sometimes called *manifest variables* in the latent-variable literature. Latent variables are not limited to conceptual variables such as intelligence, trustworthiness or decisiveness. Latent variables such as the true disease status of a patient can be measurable in principle but remain unobserved owing to the absence of an accurate method of obtaining the measurement in a particular setting.

At a more directly empirical level, there are several reasons why the use of latent variables is valuable in statistical analysis. First, the directly measured variables may be intrinsically uninteresting.

Illustation: Latent variable   Arithmetic ability may be studied by supplying to a subject a series of *items* each consisting of a question with a choice of two answers, the subject being required to mark the correct answer. It is clear that the items are of no intrinsic interest. They are worth studying only to the extent that they illuminate the respondent's arithmetical knowledge and/or ability, which for the purposes of quantitative study is a latent variable, in fact a latent random variable if a population of individuals is involved.

Other somewhat similar situations may be less extreme in that some individual items are of intrinsic interest.

Illustration: Manifest and latent variables   Health-related quality of life may be studied by questionnaires addressing the physical, social and psychological state of a patient, leading to the initial notion of a three-dimensional latent variable capturing the essence of a patient's condition. Here, however, some individual items may be of considerable intrinsic concern. An example in the case of some medical conditions is the question 'Are you confined to a wheelchair'? Even if such data are analysed using some notion of latent variables, the possible special importance of specific items needs attention.

4.7.2 Role in model formulation

The second role of latent variables is as a convenient device for model construction and occasionally model interpretation. An important application is connected with regression analysis for binary variables. Here, if Y is a binary response variable taking the values 0 and 1 and depending on one or more explanatory variables x, the most direct approach is to consider a representation of a form generalizing a linear regression:

$$P(Y = 1) = E(Y) = g(\beta x), \tag{4.1}$$

where β is a regression coefficient and $g(.)$ is a suitable function. Here and throughout $P(A)$ denotes the probability of the event A and $E(Y)$ denotes the expected value, or long-run average, of the random variable Y. The most commonly used functions are the logistic and the probit functions. These are respectively

$$P(Y = 1) = \frac{\exp(\alpha_l + \beta_l x)}{1 + \exp(\alpha_l + \beta_l x)} \qquad (4.2)$$

or, in terms of $\Phi(.)$ the standard normal integral,

$$P(Y = 1) = \Phi(\alpha_n + \beta_n x). \qquad (4.3)$$

Here α denotes the intercept determining $P(Y = 1)$ at $x = 0$ and β is a regression coefficient. Different subscripts are used in (4.2) and (4.3) because, although numerically the two functional forms are for most purposes virtually identical, the values of the corresponding parameters are not.

We will discuss the interpretation of these different parameterizations further in Section 7.1.

An alternative and older approach, still useful for motivating more complicated situations, is to suppose that there is an underlying latent random variable Ξ, sometimes called a *tolerance*, such that if and only if $\Xi \leq \alpha + \beta x$ then the binary response Y takes the value 1. The distribution of Ξ is taken to be of a simple form. Possibilities are the unit logistic density and the unit normal density, namely

$$\frac{\exp(\xi)}{\{1 + \exp(\xi)\}^2}, \qquad \text{and} \qquad \frac{1}{\sqrt{(2\pi)}} \exp\left(\frac{-\xi^2}{2}\right). \qquad (4.4)$$

With these forms the relationships (4.2) and (4.3) between the observed binary response and x are obtained. See Figure 4.2.

This representation has greater appeal if the latent tolerance variable Ξ is in principle realizable.

One use of such a representation is to motivate more elaborate models. For example, with an ordinal response it may be fruitful to use a number of cut-off points of the form $\alpha_k + \beta x$. Thus, for three cut-off levels an ordinal response model with four levels would be defined; see Figure 4.3.

Multivariate binary and similar responses can be related similarly to the multivariate normal distribution, although the intractability of that distribution limits the approach to quite a small number of dimensions.

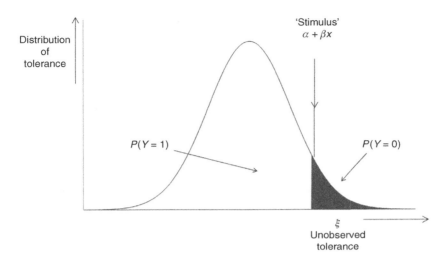

Figure 4.2 Dependence of a binary response, Y, on an explanatory variable, x. An individual has an unobserved tolerance, ξ, with distribution as shown. The stimulus $\alpha + \beta x$ depends on the explanatory variable, x. If and only if the stimulus exceeds the tolerance, the response $Y = 1$ is observed; otherwise $Y = 0$.

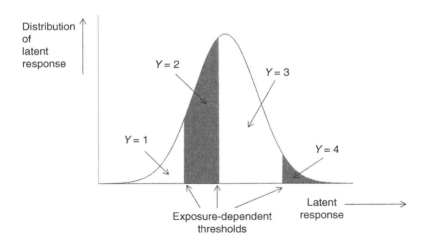

Figure 4.3 Ordinal observed value with levels $1, \ldots, 4$. The unobserved latent variable has the distribution shown. The exposure-dependent thresholds determine the observed response.

4.7.3 Latent structure and latent class models

In some contexts all or many of the features in a study are addressed by a combination of individual test scores. We have already mentioned such features as arithmetical ability and health-related quality of life. There are two broad approaches to the analysis.

One is to calculate simple summary scores for the items corresponding to each identified feature, for example the proportion of correct answers to a series of true–false questions. The notional expected value of this summary score is then taken as defining the aspect of interest and the basis of interpretation.

The second possibility is to postulate a latent variable corresponding to each feature and to regard the objective of study to be the relationship between these latent features. It is typically assumed that the observed variables are simply related to the relevant latent variable, with independent errors. There are two broad types of such relationships, latent class models and latent structure models.

In latent class models it is assumed that the study individuals each fall into one of a small number of types or classes. Within each class the relevant observations are independently distributed. The relationships between different features are expressed solely by the presumed class membership.

The simplest form of latent class analysis takes p binary variables and assumes that there are a small number k of classes and that within each class the binary variables are mutually independent. Each class thus requires the specification of p probabilities, and $k - 1$ parameters are needed to specify class membership. Thus the model involves $kp + k - 1$ parameters. This approach may be attractive if $kp + k - 1$ is appreciably less than $2^p - 1$, and especially if the latent classes can reasonably be given a subject-matter interpretation. An adequate fit may always be achieved by increasing the number, k, of latent classes but then there are clear dangers of over-interpretation.

Latent class analysis has been used to assess the accuracy of diagnostic tests when no 'gold-standard' diagnoses are available for the patients under study, only the results of multiple imperfect diagnostic tests (Pepe and Janes, 2007). In this case the true disease status is the latent variable.

Illustration: Latent class analysis   Bunting *et al.* (2008) used latent class analysis to clarify the choice of variables for the main analysis of a survey of patterns of art attendance in England. There were 729 possible responses to questions about three domains: (i) theatre, dance and cinema; (ii) music; (iii) visual arts, museums etc. For each domain, the frequency of attendance in the previous 12 months was recorded as: never; once or twice; three or more times. All possible response patterns occurred with reasonable frequency in a sample of about 30 000 individuals.

Latent class analysis suggested grouping the responses to each domain into one of three sets: Little if anything; Now and then; Enthusiastic. For cross-domain analyses the subjects were grouped into four sets: Little if anything in all three domains; Now and then in one or more domains; Enthusiastic in one domain; Enthusiastic in two or in all three domains.

These groupings were then used both for the simple tabulation of frequencies and also for studies of the relationship of art attendance to, for example, social status, education and health.

In *latent structure analysis*, by contrast, the latent variables are treated as continuous and often as normally distributed and interrelated by regression-like relationships.

4.7.4 Measurement error in regression

The last and in some ways most conceptually direct use of latent variables is to represent features that could, in principle at least, be measured but which are in fact not available. We concentrate the remainder of the discussion on this case. The most important application is to the effect, in studies of dependence, of measurement error in the explanatory variables.

We start with the simplest case, that of the linear regression of a continuous response variable Y on a continuous explanatory variable X. Random measurement error in Y will add to the scatter around the fitted regression but we will disregard that aspect to concentrate on the more subtle effects of error in X.

The latent variable here is the notional 'true' value of X, denoted X_t. The possibilities are that:

- X_t is the mean of a large number of hypothetical repetitions of the measuring process under the same conditions;
- X_t is the outcome of a 'gold-standard' method of measurement, as contrasted with the perhaps quicker and less expensive method actually employed; or

- X_t is a point value, or sometimes an average over a suitable time period, of a property varying erratically over time.

An example of the last possibility is the measurement of blood pressure. This is known to fluctuate substantially over relatively short time periods, although typically only the value at one time point will be measured and recorded.

The difference between the measured value X_m and X_t is the measurement error, and its effect on the apparent relationship between Y and X depends on a number of key features.

First, a systematic error that remained constant throughout a study would displace the regression but leave its slope unchanged. A systematic error that, say, drifted over time or was different for different sections of data would possibly produce very misleading conclusions, and steps to avoid such errors, by for example the recalibration of measuring devices, are crucial.

For the remaining discussion, we assume that the differences between the measured and true values represent random variation. There are two main possibilities. The first, often called the *classical error model*, is that for continuous measurements

$$X_m = X_t + \epsilon, \tag{4.5}$$

where ϵ is a random error that has mean zero and variance σ_ϵ^2 and, particularly importantly, that is independent of X_t.

If, however, X_t is binary, an error of measurement is a misclassification, and the corresponding assumption is that

$$P(X_m = 1 \mid X_t = 0) = \eta_0, \qquad P(X_m = 0 \mid X_t = 1) = \eta_1, \tag{4.6}$$

where the measuring procedure is characterized by the two error rates, η_0 and η_1. Here $P(A \mid B)$ denotes the probability of event A given event B. In the context of diagnostic testing and assuming that $X_t = 1$ in diseased individuals, $1 - \eta_1$ and $1 - \eta_0$ are called, respectively, the sensitivity and the specificity.

For continuous X the general effect is to flatten the regression relationship; this is called attenuation or regression dilution. In fact the least-squares slopes β_t and β_m, for true and measured X, are such that

$$\beta_m = \beta_t \frac{\sigma_t^2}{\sigma_t^2 + \sigma_\epsilon^2}. \tag{4.7}$$

Here σ_t^2 is the variance of X_t and the denominator is σ_m^2, the variance of X_m. If σ_ϵ^2 can be estimated, for example from independent replication on a

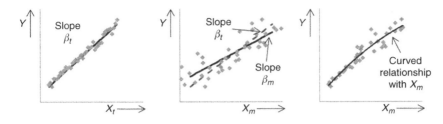

Figure 4.4 Effect of measurement error on regression. (a) Close linear relationship between Y and the true explanatory variable X_t. (b) The points in the previous plot spread out horizontally owing to the measurement error in X to give attenuated regression on the measured value X_m. (c) Proportional measurement error confines distortion to the upper end of range and induces a relationship curved at the upper end.

subsample of individuals, the attenuating factor in (4.7) can be estimated and hence an estimated slope obtained from X_m can be adjusted to provide an unbiased estimate of β_t.

An understanding of this issue is probably best obtained from some simple diagrams. Thus Figures 4.4(a), (b) show the flattening induced by random variation in the measurement of X. Figure 4.4(c) shows that if the errors are larger for large X, as for example they would be if they were proportional to X, then a change in shape is induced. This has particular implications for the exploration of, say, dose–response relationships at either extreme of a dose range.

For multiple regression with essentially independent explanatory variables, similar conclusions hold. If, however, there is strong dependence between explanatory variables then the effect of measurement error may be more confusing.

Illustration: Precision and true association in explanatory variables
An early investigation of AIDS among male homosexuals appeared to show that HIV positivity depended on the extent of use of amyl nitrite pills and not on the number of sexual partners (Vandenbroucke and Pardoel, 1989). These two potential explanatory variables were strongly associated and it is likely that the former was relatively well recorded, that is, it had only a small measurement error, whereas the latter had a large measurement error. Regression on the 'true' but badly measured variable was in effect transferred to the 'false' but well-measured correlate.

That is, the formula corresponding to (4.7) for a single explanatory variable becomes more complicated and requires knowledge of the measurement variances (and in principle covariances) for all the components.

Somewhat similar conclusions apply to other forms of regression, such as that for binary responses.

In addition to the technical statistical assumptions made in correcting for measurement error, there are two important conceptual issues. First, it is sometimes argued that the relationship of interest is that with X_m, not that with X_t, making the discussion of measurement error irrelevant for direct purposes. This may be correct, for example for predictive purposes, but only if the range of future values over which prediction is required is broadly comparable with that used to fit the regression model. Note, for example, that if in Figure 4.4(b) the values to be predicted lay predominantly in the upper part of the range then the predicted values would be systematically biased downwards by use of the regression on X_m. Second, it is important that the variation used to estimate σ_ϵ^2 and hence the correction for attenuation should be an appropriate measure of error.

Illustration: Correcting for measurement error   In INTERSALT, a large international study of the effect of salt intake on blood pressure, salt intake was estimated from analysis of urine (Stamler, 1997). For a small proportion of subjects, duplicate observations of salt intake were made a few days apart and used to estimate the variance associated with the measurement error. This correction for measurement error greatly increased the slope of the blood pressure versus sodium intake regression and enhanced the implied importance of limiting salt intake for the prevention of cardiovascular disease. There was, however, a dispute, not clearly resolved, about whether the correction for attenuation was justified; did the error variance as estimated really reflect error or did it in fact reflect true temporal variation in the salt intake and therefore in blood pressure?

The above discussion applies to the classical error model, in which the measurement error is statistically independent of the true value of X. There is a complementary situation, involving the so-called Berkson error, in which the error is independent of the measured value, that is,

$$X_t = X_m + \epsilon^*$$ (4.8)

where ϵ^* is independent of X_m, implying in particular that the true values are more variable than the measured values (Reeves *et al.*, 1998). There

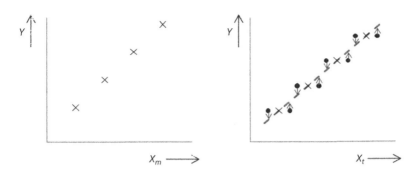

Figure 4.5 Berkson error. *Left-hand panel*, the measured values, equally spaced; *right-hand panel*, the true values (solid circles). The true values are distributed around the measured values, as shown by the arrows. The induced responses continue to lie on the regression line. There is no attenuation but increased variance about the regression line.

are two quite different situations where this may happen, one experimental and one observational. Both are represented schematically in Figure 4.5. In the experimental context a number of levels of X are pre-set, corresponding perhaps to equally spaced doses, temperatures etc. These are the measured values. The realized values deviate from the target levels by random amounts, making the formulation (4.8) appropriate.

An illustration of an observational context for Berkson error is the following.

Illustration: Berkson error   In case-control studies of the effect of exposure to radon in the home on lung cancer, the residence addresses for subjects were identified over an appropriate time period and wherever possible radon meters were placed in these homes to provide exposure measures with small error. In those cases where the home was inaccessible, exposures were imputed by averaging values over neighbouring similar sites. Under the assumption that the exposure at a particular site varies randomly from the local geographical mean and that the latter is estimated precisely, the Berkson error structure applies. If the error in the local geographical mean is appreciable then we have a situation intermediate between classical and Berkson errors.

Figure 4.5 illustrates in the case of a linear regression why the impact of Berkson error is so different from that of classical error. The induced error translates into an additional error in the response Y, averaging to zero and inducing no change in the regression relationship itself. For nonlinear relationships a change is induced.

The distinction between classical and Berkson error is thus important. The decision about which is at play has to be made on the basis of the nature of the measuring process; it cannot be made on the basis of a statistical test. The only exception is in the unlikely situation where independent estimates are available of the variances of both X_t and X_m, when, if one turns out to be clearly the greater, the broad nature of the error structure is implied.

Notes

Section 4.1. For a broad discussion of the principles of measurement see Hand (2004). For a discussion of methods for assessing health-related quality of life, see Cox *et al.* (1992).

Section 4.2. For a brief account of content analysis, see Jackson (2008).

Section 4.7. Latent structure analysis is a development of factor analysis, a method widely used in early work on educational testing. A key paper is that of Jöreskog and Goldberger (1975). For a detailed discussion of latent structure models, see Bollen (1989). Latent class analysis was introduced by Lazarsfeld in a famous study, *The American Soldier* (Merton and Lazarsfeld, 1950). For a detailed account of measurement error in regression analysis, see Carroll *et al.* (2006) and, for a brief introduction, Reeves *et al.* (1998).

5

Preliminary analysis

An outline is given of some of the steps needed to ensure that the data finally analysed are of appropriate quality. These include data auditing and data screening and the use of simple graphical and tabular preliminary analyses. No rigid boundary should be drawn between such largely informal procedures and the more formal model-based analyses that are the primary focus of statistical discussion.

5.1 Introduction

While it is always preferable to start with a thoughtful and systematic exploration of any new set of data, pressure of time may tempt those analysing such data to launch into the 'interesting' aspects straight away. With complicated data, or even just complicated data collection processes, this usually represents a false time economy as complications then come to light only at a late stage. As a result, analyses have to be rerun and results adjusted.

In this chapter we consider aspects of data auditing, data screening, data cleaning and preliminary analysis. Much of this work can be described as forms of data exploration, and as such can be regarded as belonging to a continuum that includes, at the other extreme, complex statistical analysis and modelling. Owing to the fundamental importance of data screening and cleaning, guidance on ethical statistical practice, aimed perhaps particularly at official statisticians, has included the recommendation that the data cleaning and screening procedures used should be reported in publications and testimony (American Statistical Association Committee on Professional Ethics, 1999).

Of course the detailed nature of these preliminary procedures depends on the context and, not least, on the novelty of what is involved. For example, even the use of well-established measurement techniques in the research laboratory needs some broad checks of quality control, but these requirements take on much more urgency when novel methods are used in the field by possibly inexperienced observers. Moreover, for large studies breaking new ground for the investigators, especially those requiring the development of new standardized operating procedures (SOPs), some form of pilot study is very desirable; data from this may or may not be incorporated into the data for final analysis. More generally any changes in procedure that may be inevitable in the course of a long investigation need careful documenting and may require special treatment in analysis. For example, in certain kinds of study based on official statistics key definitions may have changed over time.

5.2 Data auditing

By data auditing we mean the auditing of the process of identifying the appropriate units for data collection, the measurement of variables and the recording of data. In other contexts, data auditing is used to refer to the process of identifying all datasets held within an organization, identifying the individuals responsible for them and providing a structure to support the sharing of sets of data, as appropriate, within an organization and with others.

A data audit traces backward from the recorded data as far as reasonably possible. If the audit goes back to the original source of data, for example by remeasuring subjects or by resurveying land for features of interest, this process is sometimes referred to as 'ground truthing' the data. It may be helpful to take just a small sample of data as part of a pilot audit.

Illustration: Quality assurance audit and reanalysis   Prominent examples of large-scale data audits are the audits and reanalyses undertaken of the Harvard Six Cities Study and the American Cancer Society (ACS) Study (Krewski *et al.*, 2000, 2003). A reanalysis team undertook a quality assurance audit of a sample of the original data and sought to validate the original numerical results. The data subjected to the quality assurance audit included both data on the study population and, for the Six Cities Study, data on air quality. As to be expected in such large studies, the reanalysis found a small number of errors in coding and in the inclusion of a small number of subjects who should have been excluded. However, these errors did not affect the conclusions.

Illustration: Independent data extraction for quality assurance   An independent auditor was appointed to assess the quality of the data collected in a large field trial of badger culling policies aimed to reduce the risk of tuberculosis in cattle herds. The primary data were the number of new detections of tuberculosis in cattle herds, termed tuberculosis herd breakdowns, within the study areas. These data were extracted from a multi-disease animal health database used to record surveillance data for the government department with responsibility for animal health. The data could not be traced back to the original cattle in which evidence of tuberculosis was disclosed, because the animals involved had been slaughtered to protect public health and to avoid further spread of the disease. However, the auditor worked with staff in a government agency to extract, independently of the scientists overseeing and analysing the trial, the counts of tuberculosis herd breakdowns by area and time period. This provided an important independent check of the quality of the single most important type of data collected.

Illustration: Genetic test changes conclusion   Data auditing can be critical even for studies with a small sample size. In October 2001 the UK Department of Environment, Food and Rural Affairs announced that it had been discovered that sheep brain tissues intended for testing for bovine spongiform encephalopathy, BSE, also known as mad cow disease, had been contaminated with cattle brain tissues. Critically, the brain sample that had been found positive for BSE was in fact bovine tissue rather than ovine tissue. A positive pure sheep sample would have had major implications for the British sheep industry. Previously, in September 2001, ministers had released contingency plans that included a worst-case scenario in which 'the eating of lamb would be banned, and the entire UK flock of up to 40 million sheep would be destroyed'. There are a number of morals to this episode; one is the need for independent replication within a single investigation.

5.3 Data screening

The purpose of data screening is an initial analytical, that is, non-audit, assessment of data quality. Questions of data quality must, however, be kept in mind throughout the process of data analysis and interpretation, because complex issues may well not be discovered until considerable analysis has been undertaken.

The first thing to explore is the pedigree of the data. How were the data collected? How were they entered into the current electronic dataset or database? Were many people involved in data collection and/or data entry? If so, were guidelines agreed and recorded? How were dates recorded and entered? After all, 05/06/2003 means 5 June to some and May 6 to others.

A starting point in dealing with any set of data is a clear understanding of the data coding structure and of any units associated with the variables. Coding any missing values as −99 may not be confusing if the variable in question is age in years. However, coding them as 'Unk' when the variable denotes names of individuals, for example, of animals, may be confusing particularly when many individuals have multiple entries in the dataset. Thus, 'Unk' might be interpreted incorrectly as the name, albeit an unusual one, of an individual.

Calculations of means, standard deviations, minima and maxima for all the quantitative variables allow plausibility checks and out-of-range checks to be made. An error could be as simple as a misinterpretation of units, pounds being mistaken for kilograms, for example. Input and computational errors may become apparent at this stage.

It may help to consider multiple variables simultaneously when searching for outliers (outlying individuals), the variables being selected if possible only after detailed consideration of the form in which the data were originally recorded. For example, it may become clear that height and weight variables were switched at data entry. Furthermore, from a plot of two variables that are strongly related it may be easy to detect outlying individuals that are not anomalous in either variable on its own. Such individuals may have a major, often unwanted, effect on the study of relationships between variables.

Illustration: Spreadsheet errors　Powell *et al.* (2009) examined spreadsheets to characterize the frequency and types of error that arise during their use. On this basis they defined six error types: logical errors, where a formula was used incorrectly; reference errors, where a formula refers to another cell incorrectly; the placing of numbers in a formula, not necessarily an error but considered bad practice; copy and paste errors; data input errors; and omission errors, where an input cell required by a formula is missing. They estimated that 1% to 2% of all formula cells in spreadsheets contain errors. As many spreadsheets contain a large number of formula cells, the potential for error is considerable. Careful inspection of the original spreadsheet(s) or program commands is clearly warranted whenever possible if the data were processed in any way following data entry.

Identifying univariate outliers may in a few cases be relatively simple, whereas detecting multivariate outliers usually requires more work. Outliers will affect the descriptive statistics such as sample means and variances, so it is possible that outliers may be obscured when simple rules based on these quantities are used to identify them. More challenging, in some respects, than the identification of outliers is determining what to do about them. Outliers that arise from errors should be excluded if they can be thus identified conclusively. However, outliers may arise from rare but accurate observations which could provide important insight into the system under study. In the investigation of outliers, subject-matter expertise is important in identifying observations that go beyond what would have reasonably been expected. Judging exactly how rare or implausible a particular sort of outlier is may be particularly difficult when previously published analyses have excluded such observations and have presented little or no detail on them.

In some situations with very large amounts of data, a small proportion of which are subject to large errors of measurement, it is important to use methods that are automatically insensitive to such extremes.

Illustration: Outlier detection to detect foul play   In some settings, the detection of outliers may be one main purpose of the analysis. An example of this is financial fraud detection based on the detection of anomalous transaction data and/or user behaviour (Edge and Sampaio, 2009). Similarly, possible intrusions into or misuses of computer systems can be detected by analysing computer use patterns and assuming that instances of computer misuse are both rare and different from patterns of legitimate computer use (Wu and Banzhaf, 2010).

Illustration: Data exclusion to avoid undue influence of irrelevant factors   Data screening may lead to some data being excluded not because they were errors but because they were dominantly affected by aspects other than the phenomenon under study. A study of aerosol particles detected in Chichi-jima of the Ogasawara Islands in the northwestern

Pacific aimed to characterize the black-carbon mass fractions in anthropogenic aerosols from the source regions, China, Japan and the Korean peninsula (Koga *et al.*, 2008). However, because aerosol concentrations are strongly affected by rain, particle data were excluded during time periods when local precipitation records indicated that it had rained. Furthermore, the coefficient of variation, that is, the standard deviation divided by the mean, in the concentrations was calculated for four particle-size classes for each hour. If the coefficient of variation was greater than one-half in one or more of the size classes then all data from that time interval were excluded to remove the influence of local sources of carbon particles.

Patterns of missing data should be explored at an early stage to ensure that any potential problems connected with missingness are identified promptly. The first assessment should examine on a univariate basis the proportion of data which are missing for each variable. If a particular variable, not the strong focus of the investigation, has a high proportion of missing data then consideration should be given to whether the variable could or should be dropped from further analyses. However, a rule such as 'drop the variable if more than 25% of values are missing' should never be applied blindly. Furthermore, in some situations variables may have data missing by design, so-called planned missingness.

Illustration: Planned missingness   Planned missingness might arise if a questionnaire were divided into sections and respondents were only asked questions from some sections of the questionnaire (Raghunathan and Grizzle, 1995). Alternatively, a two-stage design would give rise to planned missingness if the first stage involved the testing of all subjects with a relatively cheap diagnostic test while the second stage involved testing only a subset of these subjects with a more expensive test. See for example McNamee (2002).

Further investigations into correlates of missing values, or outliers, could include logistic regression models to predict which individuals will be missing or will have an outlier value for a particular variable. Depending on the amount of missing data and the extent to which it is non-random, techniques such as expectation-maximization (EM) imputation may be appropriate to obtain estimates that make effective use of all the available data. Indeed, virtually all the more elaborate methods of imputing missing values assume that missingness is random and in particular that it is conditionally independent of the value that would have been observed if complete data collection had been achieved. Techniques for dealing with so-called informative missingness depend on strong and typically untestable assumptions.

Illustration: Exploring patterns of data missingness Patterns of missingness in multiple variables may be explored using standard descriptive techniques. Data on health-related quality of life (QOL) were collected from a melanoma trial that compared two treatments (Stubbendick and Ibrahim, 2003). Once patients enrolled in the trial they were asked to complete QOL questionnaires at baseline, after one month, after six months and after one year. Of the 364 patients enrolled, 54 cases died before all four measurements could be taken and 33 cases had no QOL data at any time point. From the remaining 277 patients, 11% had data missing at baseline, 17% at one month, 22% at six months and 27% at one year. Unsurprisingly, even among the 277 with some QOL data, the missing data were clustered with 159 patients having complete QOL data and 34 patients having only one of the four QOL measures complete, which may be compared with the expected values 118 and 6 respectively, had data been missing at random.

Pairwise relationships between variables should be assessed to determine whether the variables are too highly correlated for some purposes. For example, from a set of highly correlated variables a single variable may need to be chosen for inclusion as a predictor in a regression model, the others being regarded as, in a sense, redundant. If such exclusions are not made, the precision of regression slopes is likely to be reduced and, more importantly, individual estimates of slope may be misleading. In contrast, all variables regardless of collinearity may be included in a factor analysis without any ill effects. This is discussed further in Section 7.3.

In some situations the pattern and amount of random variability may vary either haphazardly or systematically, and this may be both of intrinsic concern and have important implications for the subsequent analysis.

For studies in which relatively little is known about the spatial distribution of an outcome of interest, there is a range of exploratory techniques available, the use of which is sometimes referred to as exploratory spatial data analysis. Depending upon the specifics of the study at hand, aims may include the identification of hot spots (clusters) and cool spots, a particularly treacherous issue in epidemiology. Somewhat easier to investigate are systematic spatial associations, the detection of spatial outliers and the identification of sources of spatial heterogeneity. The characterization of spatial heterogeneity may be a central aim of the project or, at the other extreme, a nuisance required to avoid an overestimation of precision if account has not been taken of substantial spatial correlation.

5.4 Preliminary graphical analysis

A wide range of graphical tools for the description and exploration of data is easily accessible. A histogram shows the distribution of a variable, whereas a boxplot gives in more compact form the median, skewness and any outliers in the distribution. Boxplots may also be a convenient way of illustrating comparisons between a small number of datasets. Maps also may be important tools. New data visualization tools have been developed within geographic information systems (GISs) to assess patterns hidden within large spatial sets of data, for example those collected from satellites. Dynamic graphical representations are likely to be particularly helpful for studying both empirical data and the output of complex computer models where the temporal or spatial development of processes is involved.

Illustration: Graphical demonstration of association   A famous early graphical display of data was the map used by John Snow to show the locations of cholera deaths and of sources of drinking water on the same London city map in 1854 (Snow, 1855). See Figure 5.1. Historical research indicates that rather than using the map to discover the likely source of the epidemic, Snow used the map to convey the conclusions he had made on the basis of his theory of how cholera was transmitted (Brody *et al.*, 2000). Snow had seen the map published by an Exeter doctor, Thomas Shapter, showing the locations of cholera deaths on a map of Exeter; however, this map crucially did not show the locations of sources of drinking water (Shapter, 1849).

Illustration: Graphical depiction of complex data structure   Graphical representation can be particularly helpful in understanding networks. A study was undertaken to describe a cluster of HIV transmission in South Wales (Knapper *et al.*, 2008). The identification and notification of partners following diagnosis of the index case allowed reconstruction of the sexual network through which infection was spread. While a plot of the

Figure 5.1 Part of a map of cholera deaths, showing sources of drinking water (Snow, 1855).

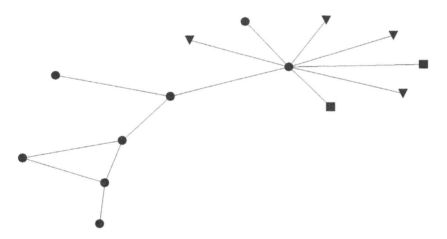

Figure 5.2 Plot of a sexual contact network in which individuals are plotted as nodes and sexual contacts are plotted as edges. Plotting symbols are used to distinguish the different attributes of the nodes.

network using individuals as nodes and sexual contacts as edges provides insights, further information is conveyed through the use of different shapes and shading to distinguish attributes of the nodes (as in Figure 5.2).

A scatterplot describes a bivariate distribution and allows bivariate outliers to be identified visually. A scatterplot matrix is a multi-panel display presented in much the same way as a correlation matrix, in which each column relates to the same x-axis and each row to the same y-axis. Thus, for the variables included, a scatterplot matrix presents all the bivariate relationships, each combination appearing both above and below the diagonal owing to symmetry. While a traditional single-panelled two-dimensional scatterplot displays only two variables, additional information can be added to reflect other variables through varying the size, shape or colour of the datapoints or even through the addition of a 'tail' to each data point in which the angle of the tail reflects an additional variable. The term 'glyph plot' is used for graphics including such glyphs, defined as symbols which display data by changing their appearance. Weather maps frequently use such symbols to show wind speed and wind direction simultaneously. Their general use for statistical analysis is probably restricted to situations with a small number of distinct points on the graph. For example, it is reasonable to show intercountry relationships between pairs of features by plotting a point for each country, provided that the number of countries involved is not too large.

Some of the most rapidly developing aspects of graphical methods centre on the use of powerful computer-based techniques, in particular to examine dynamic high-dimensional dependencies. At a more primitive level, important aspects of traditional graphical methods are often disregarded, both in preliminary analysis and also in presenting conclusions. These include the following:

- axes should be clearly labelled;
- related graphs should normally be on the same scales to ensure comparability;
- false origins should be marked by a scale break;
- so far as is feasible, distinct points should be of roughly equal precision;
- so far as is feasible, distinct points should have independent sampling or measurement errors;
- while some indication of the precision of points on a plot is desirable, the provision of large numbers of confidence intervals and/or significance tests may invite misinterpretation;
- the interpretation of plots with substantial noise should not be compromised by imposing on them prominent smooth curves; and
- so far as is feasible, the legend attached to the diagram should make the meaning self-explanatory without reference back to the text.

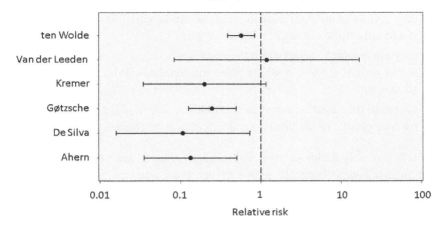

Figure 5.3 A forest plot comparing the results of six randomized controlled trials (Ahern *et al.*, 1984; De Silva and Hazleman, 1981; Gøtzsche *et al.*, 1996; Kremer *et al.*, 1987; ten Wolde *et al.*, 1996; Van der Leeden *et al.*, 1986). Four trials showed that patients who remained on their initial dosage of disease-modifying antirheumatic drugs were at significantly lower risk of flares or worsening of their rheumatoid arthritis than patients who discontinued treatment, as reviewed by O'Mahony *et al.* (2010).

One implication of these points is that there is at most a limited and specialized role for smoothing methods other than simple binning. The reason is that smoothness which is artefactually generated is virtually impossible to distinguish from real smoothness implicit in the data. Put differently, undersmoothed data are easily smoothed by eye, or more formally, but oversmoothed data are vulnerable to misinterpretation and can be unsmoothed only, if at all, by delicate analysis.

A common method, in some epidemiological studies, of presenting the analysis of a series of related investigations, sometimes called a *forest plot*, is less than ideal in some respects. Usually the primary outcome plotted is a log relative risk, that is, the log of the ratios of the estimated probabilities of death for individuals exposed or unexposed to a particular risk factor. Figure 5.3 shows a typical forest plot, in which each study provides an estimate and a 95% confidence interval.

Imperfections in this are as follows:

- confidence intervals are appropriate for assessing uncertainty in a single estimate but are less so for comparative purposes;

- they invite the misinterpretation that two studies are mutually consistent if and only if the confidence bands overlap;
- they misleadingly suggest that the studies in a set are mutually consistent if and only if there is a single point intersecting all the study intervals shown; and
- to obtain the standard error and confidence intervals for comparison of pairs or groups of studies is not as direct as it might be.

It is probably better to show the standard error of each estimate. There are further issues connected with such summary analyses, in that quite often the different estimates are adjusted for other risk factors in a way that differs between studies. Also, care is needed is indicating the precision of the overall assessment; some methods of analysis lead to the serious underestimation of potential errors in the final summary.

5.5 Preliminary tabular analysis

Tabular analysis has a long history. Indeed, such analysis was the topic of a paper read to the Royal Statistical Society in 1879 (Guy, 1879), referring to an 1831 book on the same topic (Todd, 1831). Simple descriptive tables typically gave counts, possibly accompanied by means, for a particular combination of attributes.

Illustration: Early tabular analysis   Tables were published in 1841 of 134 cases of typhoid fever in Paris, stratifying them by symptoms and severity (Walshe, 1841a,b,c). The author found the tables 'of considerable utility, in affording a ready means of comparing the symptomatology of the form of the malady existing in the French capital, and that prevailing in other quarters. Believing it possible that this analysis might similarly serve the purpose of others, I venture to publish it'.

Although such methods are sometimes considered uninteresting, owing to their relative simplicity, simple descriptive tables can play important roles in describing a sample under study. They can provide key data for subsequent analyses seeking a synthesis of different but related results from distinct studies. Furthermore, they can demonstrate the extent to which a study's conclusions were well grounded in the observed data.

Illustration: Tabular analysis demonstrates some results were off support
Messer *et al.* (2010) explored the effects of neighbourhood characteristics (economic deprivation and racial segregation) on the risk of preterm
birth. Their tabulation of women and preterm births by every combination of level of economic deprivation and racial segregation and by
county, race and maternal education level demonstrated how unevenly
the data were distributed, some combinations of economic deprivation and racial segregation being entirely absent. On the basis of these
tables, the authors concluded that their logistic regression results were
'off support' (Manski, 1993), in that they involved extrapolation to predict the risk of preterm births for groups of women for whom no data
were available. Although not necessarily to be avoided, the interpretation of off-support results should be particularly cautious.

The primary reporting of the conclusions of statistical analysis is by descriptive statistics and by estimates of parameters together with standard
errors, sometimes accompanied by standard errors of associated contrasts.
Unnecessary digits hinder the reader; at the same time a future user of the
estimates should not be faced with a loss of accuracy due to appreciable
rounding errors. Some broad guidelines based on those considerations are
as follows.

Units should always be stated and chosen so that, as far as is feasible,
estimates are neither very large numerically nor very small. For physical
measurements this can be achieved within the SI system (Le Système International d'Unités); for example, for mass the following units can be used:
..., μg, mg, g, kg, ... Standard errors should be given to two or at most
three working digits and primary estimates should have a rounding error
of less than one-tenth of a standard error. For example, a mass might be
quoted as 0.453 kg with a standard error of 0.026 kg.

5.6 More specialized measurement

Much laboratory-based work, which years ago would have largely used
apparatus built in the local workshop, now uses intricate externally manu-
factured equipment, quite often with built-in computerized analysis giving
key summaries or graphical displays and allowing, where appropriate, direct entry into a data base for a whole study. These changes do not, however, diminish the need for quality control of the data, even if the detailed
procedures may take different forms.

Perhaps the most difficult aspect to assess is that a large amount of pre-liminary processing may have taken place before the primary data become available for analysis.

Illustration: Particle physics data   The primary data for analysis from an experiment on the large hadron collider at CERN, the European Orga-nization for Nuclear Research, are counts of collisions grouped into a large number of bins defined by the energy involved. A background of occurrences arises in Poisson processes, that is, completely randomly, at a rate that varies smoothly with energy. Above this background, occurring in narrow energy bands are apparently high counts, which may represent:

- known phenomena of no immediate concern;
- occurrences expected on theoretical grounds but not yet observed, most notably an appearance of the Higgs boson;
- unexpected occurrences, so-called new physics;
- chance fluctuations to be expected in view of the very large number of energy bands studied.

The data for analysis are, however, the product of an elaborate screening procedure in which extremely large numbers of irrelevant events are removed, and this may be a data quality aspect of concern. For a general review of statistical problems in particle physics, see Lyons (2008).

In other contexts where the evaluation and scoring of complex data are involved, blind scoring of a subsample by independent experts is desirable. When data are collected over a substantial time period, calibration checks are important and also checks for sticking instruments, that is, instruments that repeatedly record the same value, often at the extreme of the feasible range for the instrument concerned. Thus a long sequence of zero rain-falls recorded in a tipping-bucket rain gauge may be strong evidence of a defective gauge.

5.7 Discussion

It is impossible to draw a clear line between analyses which are exploratory and those which form the main body of an analytical study. Diagnostic investigations into the fit of a particular model may lead back to further exploratory analyses or even to further data screening and cleaning. The process is usually, and indeed should be, iterative as further insights are gained both into the phenomenon under study and into other processes which have contributed to the generation of the observed data.

The documentation of methods and results is important throughout to avoid confusion at later stages and in publications about the work. More generally, for data of an especially expensive kind and for data likely to be of broad subsequent interest, early consideration should be given to the subsequent archiving of information. Material to be recorded should include: the raw data, not summaries such as means and variances; a clear record of the data collection methods and definitions of variable coding schemes (including the coding of missing values); any computer programs used to retrieve the study data from a complex database.

6

Model formulation

More formal methods of statistical analysis are based on a probability model for the data. This represents in idealized form the main features of the variability encountered and possibly also summarizes the data-generating process. Such models contain parameters some of which encapsulate the research questions of concern. The main aspects of probability models are reviewed and simple examples are given.

6.1 Preliminaries

Simple methods of graphical and tabular analysis are of great value. They are essential in the preliminary checking of data quality and in some cases may lead to clear and convincing explanations. They play a role too in presenting the conclusions even of quite complex analyses. In many contexts it is desirable that the conclusions of an analysis can be regarded, in part at least, as summary descriptions of the data as well as interpretable in terms of a probability model.

Nevertheless careful analysis often hinges on the use of an explicit probability model for the data. Such models have a number of aspects:

- they may encapsulate research questions and hypotheses in compact and clear form via parameters of interest, or they may specify a simple structure, deviations from which can be isolated and studied in detail;
- they provide a way of specifying the uncertainty in conclusions;
- they formalize the discounting of features that are in a sense accidents of the specific dataset under analysis;
- they may represent special features of the data collection process;
- they allow the comparison of different methods of analysis and, in particular, they specify methods of analysis that are in a well-defined sense efficient under the conditions postulated.

In connection with this last point note that, while it is appealing to use methods that are in a reasonable sense fully efficient, that is, extract all relevant information in the data, nevertheless any such notion is within the framework of an assumed model. Ideally, methods should have this efficiency property while preserving good behaviour (especially stability of interpretation) when the model is perturbed.

Essentially a model translates a subject-matter question into a mathematical or statistical one and, if that translation is seriously defective, the analysis will address a wrong or inappropriate question, an ultimate sin.

The very word 'model' implies that an idealized representation is involved. It may be argued that it is rarely possible to think about complex situations without some element of simplification and in that sense models of some sort are ubiquitous. Here by a model we always mean a probability model.

Most discussion in this book concerns the role of probability models as an aid to the interpretation of data. A related but somewhat different use of such models is to provide a theoretical basis for studying a phenomenon. Furthermore, in certain cases they can be used to predict consequences in situations for which there is very little or no direct empirical data. The illustration concerning *Sordaria* discussed below in Section 6.3 is a simple example of a model built to understand a biological phenomenon. In yet other contexts the role of a probability model may be to adjust for special features of the data collection process.

Stereology is concerned with inferring the properties of structures in several dimensions from data on probes in a lower number of dimensions; see Baddeley and Jensen (2005). In particular, many medical applications involve inferring properties of three-dimensional structures in the body from scans producing two-dimensional cross-sections.

Often, more detailed models involve progression in time expressed by differential equations or progression in discrete time expressed by difference equations. Sometimes these are deterministic, that is, they do not involve probability directly, and provide guidance on the systematic variation to be expected. Such initially deterministic models may have a random component attached as an empirical representation of unexplained variation. Other models are intrinsically probabilistic. Data-generating processes nearly always evolve over time, and in physics this manifests itself in the near-ubiquity of differential equations. In other fields essentially the same idea may appear in representations in terms of a sequence of empirical dependencies that may be suggestive of a data-generating process.

More formal detailed models incorporating specific subject-matter considerations can be classified roughly as either 'toy' models, where the word 'toy' is not to be taken dismissively, or as quasi-realistic. Toy models are used to gain insight into possibly quite complex situations, such as epidemics, by concentration on a few key features. Such models are often best studied by mathematical analysis, although a purely numerical approach is always available. Quasi-realistic models virtually always demand computer simulation. Sometimes a combination of the two approaches may be effective. The output from a quasi-realistic model may be compared with that from the nearest toy model. This smoothes the output from the simulation and exposes those conditions under which the toy model is adequate and those in which it is seriously inadequate. This latter information may itself be enlightening. The greatest difficulty with quasi-realistic models is likely to be that they require numerical specification of features for some of which there is very little or no empirical information. Sensitivity analysis is then particularly important.

Illustration: Parallel development of 'toy' and quasi-realistic models   To analyse data on the 2003 Severe Acute Respiratory Syndrome (SARS) epidemic in Hong Kong, Riley *et al.* (2003) created a stochastic transmission model in which the population of Hong Kong was stratified by infection status (susceptible, latently infected, infectious, hospitalized, recovered or dead) which could vary over time. The population was also stratified by geographic district, and this was assumed fixed over the time period of analysis. During the analysis, the authors developed and analysed in parallel the above complex transmission model and a simplified and analytically tractable differential-equation-based transmission model. The agreement obtained between the results of these two approaches, when they could reasonably be compared, helped in checking the computer code and provided reassurance that the complex and hopefully more realistic model was giving useful results.

We suppose in most of our discussion that a specific set of data is available for analysis. Of course, this presupposes that certain choices have been made already.

6.2 Nature of probability models

It is first useful to distinguish conceptually two broad situations. In one, ideally data could be obtained largely free of haphazard variation and the aspects of interest studied directly but in fact perturbations of no intrinsic interest distort the position; such variability, which may be called *error*, needs consideration. This is the position in some laboratory investigations, especially in the physical sciences. The other possibility is that there is substantial natural variability in the phenomenon under study. In some statistical writing this variability too is called error, but this is potentially misleading.

Illustration: Variation may or may not be of intrinsic interest   Repeat determinations of the rate constant of a well-defined chemical reaction may show relatively small variability around the mean. Such variability may reasonably be called error and may need assessment if refined estimation is required of contrasts between the rates in slightly different set-ups. Care to avoid appreciable systematic error is crucial. The distribution of the random part of the error is normally of no intrinsic interest but, say, if laboratory mice are infected with a pathogen then their time to death is likely to have an appreciably dispersed distribution and the characterization of this distribution is of intrinsic interest; typically it makes no sense to talk of *the* time to death. For contrasts, say of different pathogens, any systematic change in the distribution will be of interest. It may be very helpful to describe the distribution in terms of a small number of parameters, contrasts of which can then be studied, but even then it is the distribution itself that is of ultimate concern.

In one broad class of situations we have data on study individuals for each of which a number of features are recorded. That is, there is repetition of the same or a very similar process across individuals. In other contexts the repetition may be across time, across space or across space and time. The different features typically serve different purposes, as discussed in Chapter 4.

Model formulation may proceed in a number of stages but a first step is often to consider how certain outcome or response variables depend on the explanatory variables. This translates in probabilistic terms into studying the conditional distribution of a variable Y given the values of explanatory variables x. Typically both Y and x are vectors. For this we produce a representation of the conditional distribution of Y given x, written in the form $f_Y(y, x; \theta)$. Here θ is a vector of unknown constants.

Thus a first question in model formulation concerns what is to be treated as random and what as fixed, that is, taken conditionally for the specific analysis. This is to a large extent a subject-matter question, not something to be decided by a statistical procedure.

We denote by Y the set of observations to be represented by random variables in a probability model. A model then specifies the distribution of Y. In a few special cases the model will contain a single completely known probability distribution.

Illustration: Probability distribution with no unknown parameters   A number n of organisms are released in a supposedly homogeneous environment and their direction of first movement is recorded as an angle in $(0, 2\pi)$ relative to a defined reference direction. The initial probability model is that the vector of n observations has the distribution of n mutually independent random variables uniform on $(0, 2\pi)$. Here the research question of interest concerns the consistency of the data with the specified model. Depending on the context, and possibly influenced by preliminary inspection of data, we might look for concentration of the first-movement directions around a direction given *a priori*, or around an unknown direction or around several specific directions or possibly for the avoidance of a particular set of directions. Alternatively, we might look to see whether organisms tended to follow or avoid the paths of organisms tested before them.

Illustration: Model tailored to a specific research question   Suppose that observations are obtained, on a sample of individuals, of systolic blood pressure, body mass index (BMI), age and gender and that the research question of interest concerns the dependence of systolic blood pressure on BMI for individuals of a given age and gender. All four variables vary across individuals and in different contexts all could be modelled as random variables. For the present specific purpose, however, only the systolic blood pressure is regarded as a random variable. If we denote the four variables by y and $x = (x_1, x_2, x_3)$ respectively, a model could take various forms but a simple version would specify the dependence of $E(Y)$ on the explanatory variables x, perhaps by a linear function. A second and important aspect of such a model concerns the assumption of statistical dependence or independence of the data for distinct individuals.

If the research question of interest changes and now we are interested in the relationship between BMI and age the formulation changes and the BMI is regarded explicitly as a random variable.

Much more commonly, however, we start not with a single distribution but with a family of possible distributions for Y. We write for the family of possible probability distributions the expression $f_Y(y; \theta)$, where each possible value of θ specifies a particular distribution. Here to simplify the notation we omit the dependence on any explanatory variables x. An initial broad classification of such formulations is:

- parametric;
- nonparametric; or
- semiparametric.

We will concentrate largely on parametric formulations in which θ has a finite, sometimes relatively small, number of components. Familiar examples are the standard forms of generalized linear regression, in particular those with independent normally distributed deviations; in this case the parameter θ consists of the unknown regression coefficients and the variance of the associated normal distribution. One of the simplest such forms is a straight-line relationship between the explanatory variables x_j and the response variables Y_j, namely

$$Y_j = \beta_0 + \beta_1 x_j + \epsilon_j, \tag{6.1}$$

where for $j = 1, \ldots, n$ the deviations ϵ_j are independently normally distributed with mean zero and unknown variance σ^2 and each value of j corresponds to a distinct study individual. The vector parameter is $\theta = (\beta_0, \beta_1, \sigma^2)$, although in particular instances one or more of the components might be known. The systematic component of the model can be written as

$$E(Y_j) = \beta_0 + \beta_1 x_j. \tag{6.2}$$

Many widely used models are essentially generalizations of this.

By contrast a nonparametric formulation of such a relationship might be

$$Y_j = \phi(x_j) + \epsilon_j, \tag{6.3}$$

where $\phi(x)$ is an unknown function of x constrained only by some smoothness conditions or by being monotonic, and the ϵ_j are mutually independent random variables with median zero and with otherwise unknown and arbitrary distribution.

There are two broad forms of semiparametric formulation. In one the distribution of the ϵ_j remains nonparametric whereas the regression function is linear or has some other simply parameterized form. In the other the roles are reversed. The function $\phi(x)$ remains as in the nonparametric form but the ϵ_j are assumed, for example, to be independently normally distributed with constant variance.

In this book we concentrate mostly on parametric formulations, although many of the points to be discussed apply quite broadly. Parametric models typically represent some notion of smoothness; their danger is that particular representations of that smoothness may have strong and unfortunate implications. This difficulty is covered for the most part by informal checking that the primary conclusions do not depend critically on the precise form of parametric representation. To some extent such considerations can be formalized but in the last analysis some element of judgement cannot be avoided.

One general consideration that is sometimes helpful is the following. If an issue can be addressed nonparametrically then it will often be better to tackle it parametrically; however, if it cannot be resolved nonparametrically then it is usually dangerous to resolve it parametrically.

Illustration: Survival analysis with two modes of failure   An industrial component may fail in one of two quite different failure modes. A person in remission after treatment for cancer may have a local recurrence or may have a widely dispersed recurrence. Such situations are commonly modelled by supposing that for each individual there are two notional failure times, T_1, T_2, corresponding to the two modes and that all that is observed is the smaller of the two times, $\min(T_1, T_2)$, and an indicator of which mode occurred. These questions might arise. Are T_1 and T_2 statistically independent? What is the marginal distribution of, say, T_2? It can be shown that the first question cannot be answered nonparametrically: any possible set of data is consistent with the independence of T_1 and T_2. Yet it could be answered parametrically, for example by assuming (T_1, T_2) to have a bivariate log normal distribution. The resulting estimate of the correlation coefficient between $\log T_1$ and $\log T_2$ is then extremely sensitive to the assumption of log normality and it would be wise to use such a resolution only exceptionally.

The word 'model' implies that a representation is at best an idealization of a possibly complicated real physical, biological or social system. Once a model is formulated two types of question arise. How can the unknown parameters in the model best be estimated? Is there evidence that the model needs modification or indeed should be abandoned in favour of some different representation? The second question is to be interpreted not as asking whether the model is true but whether there is clear evidence of a specific kind of departure implying a need to change the model so as to avoid distortion of the final conclusions.

More detailed discussion of the choice of parameters is deferred to Section 7.1, and for the moment we make a distinction only between the parameters of interest and the nuisance parameters. The former address directly the research question of concern, whereas typically the values of the nuisance parameters are not of much direct concern; these parameters are needed merely to complete the specification.

In the linear regression model (6.1) the parameter β_1, determining the slope of the regression line, would often be the parameter of interest. There are other possibilities, however. For example interest might lie in the intercept of the line at $x = 0$, i.e. in β_0, in particular in whether the line passes through the origin. Yet another possible focus of interest is the value of x at which the expected response takes a preassigned value y^*, namely $(y^* - \beta_0)/\beta_1$.

6.3 Types of model

Models can be classified in many ways, of which probably the most important is by the extent to which they are either *substantive* or *purely empirical*. Substantive models contain aspects that are specific to the underlying subject-matter, for example being derived from a quantitative theory of the subject-matter. Some of these models are directly probabilistic whereas others essentially consist of a deterministic theory with a perhaps largely empirical representation of random variability superimposed on it.

To some extent the two types of model correspond to different types of objective; in many contexts both may be involved at different stages. The first objective is to establish as securely as reasonably possible the existence, or sometimes the complete absence, of particular kinds of dependence. Ideally this involves the comparison of randomized treatments under clearly specified conditions with clearly defined response variables. The complementary task is to aim for understanding of the underlying processes and this tends to be intrinsically more speculative. The general ethos of most statistical discussion is cautious; this is usually appropriate and fruitful in the first kind of objective but very much less so in the second.

Purely empirical models have very little or no specific subject-matter base. They represent patterns of variability that commonly arise and indeed often derive their importance from their wide applicability to many subject-matter fields. Many methods described in textbooks on statistical

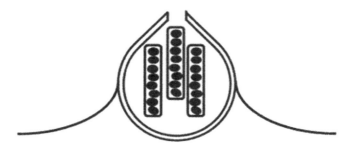

Figure 6.1 A schematic cross-section of the organism *Sordaria*.
Each group contains eight spores.

methods and implemented in widely available statistical software are of
this kind.

Illustration: Probability model directly representing underlying biological
mechanism   *Sordaria* is a small fungus containing large numbers of
spores arranged in sets of eight spores (Ingold and Hadland, 1959). See
Figure 6.1. As each set is ejected from the organism the bonds join-
ing the spores may or may not break, so that each set may generate
one group of eight spores if none of the bonds breaks or eight separate
groups of a single spore if all seven bonds break or any other partition of
eight into components. Data are collected after a large number of spores
have been ejected, giving the proportion of groups of eight, seven, . . . ,
single spores. The proportions take on stable forms as more and more
observations are taken, leading to a probability model in which π_j is the
probability that a group is of size j for $j = 1, \ldots, 8$, where $\Sigma \pi_j = 1$.

   This model merely specifies an arbitrary set of marginal probabilities. It contains no specific subject information and is a purely empirical
representation, potentially appropriate for any set of observations each
one of which may take on any of eight possible forms. Note that in
fact each primary observation is now a group of spores, not the original
set of eight which may have formed one group or many groups. Because of the special way in which the data have been generated, involving relatively subtle dependencies between the different frequencies, the
multinomial distribution over eight cells is not an appropriate model for
the full set of observed frequencies. At this stage the model is purely
empirical.

We may, however, form a substantive model for this situation that represents the data-generating process. Suppose that, on ejection of each set, each bond breaks with unknown probability θ and that all bonds and all sets are mutually independent. Then, for example, the probability that the first two bonds break and the remainder survive is $\theta^2(1 - \theta)^5$; this generates three groups, two with a single spore and one with six spores. The enumeration of all cases, combined with a specification of how the data are defined, generates the following set of probabilities:

$$\pi_1 = 2\theta(1 + 3\theta)/(1 + 7\theta),$$
$$\pi_2 = 2\theta(1 - \theta)(1 + \tfrac{5}{2}\theta)/(1 + 7\theta),$$
$$\pi_3 = 2\theta(1 - \theta)^2(1 + 2\theta)/(1 + 7\theta),$$
$$\pi_4 = 2\theta(1 - \theta)^3(1 + \tfrac{3}{2}\theta)/(1 + 7\theta),$$
$$\pi_5 = 2\theta(1 - \theta)^4(1 + \theta)/(1 + 7\theta), \tag{6.4}$$
$$\pi_6 = 2\theta(1 - \theta)^5(1 + \tfrac{1}{2}\theta)/(1 + 7\theta),$$
$$\pi_7 = 2\theta(1 - \theta)^6/(1 + 7\theta),$$
$$\pi_8 = (1 - \theta)^7/(1 + 7\theta).$$

This now leads to the technical statistical problems, not addressed in this book, of estimating θ and of examining the adequacy of the model. However, Figure 6.2 demonstrates the fit of the model to the observed data.

Illustration: One- and two-hit models   A slightly less specific example is that of one- and two-hit models for the dependence of a binary response on an explanatory variable x. In this the probability of, say, a positive outcome is $1 - e^{-\theta x}$ in the one-hit model or $1 - (1 + \theta x)e^{-\theta x}$ in the two-hit model. The motivation is that of an individual exposed for a period x to a Poisson process of point events of unknown rate θ. The individual has a positive response if it experiences at least one point event in the Poisson process, in the first case, or at least two point events, in the second. This, the Armitage–Doll model, was suggested as a representation of cancer incidence (Knudson, 2001).

The two illustrations above of substantive models are both stochastic, that is, they are essentially probabilistic in that they aim to represent the random variation in the data as well as, in the second case, the systematic dependence on the explanatory variable x. A relatively common type of

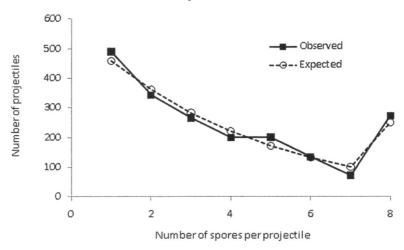

Figure 6.2 Observed and fitted frequencies for the numbers of *Sordaria* spores per projectile in an experiment conducted at 21–24 °C (Ingold and Hadland, 1959).

substantive model is based on a deterministic account of the systematic component of the variation, quite often in the form of differential equations. This is then combined with a relatively *ad hoc* empirical representation of the haphazard component.

Illustration: Compartmental transmission model   Important stochastic models that are more complicated, in particular by being dynamic, arise in studying epidemics. For example, Forrester *et al.* (2007) modelled the epidemic process of *Staphylococcus aureus* in an intensive care unit by treating patient colonizations as stochastic events, other key events (hospital admission, isolation of patients and hospital discharge) treated as specified and not modelled probabilistically (Figure 6.3). Imperfect diagnostic tests for the presence of *Staphylococcus aureus* meant that the process was only partially observable, even if patients were frequently tested for colonization.

Illustration: Michaelis–Menten equation   The Michaelis–Menten equation is widely used in biochemistry. It connects the rate of a reaction in which substrate is converted into a product with an enzyme as a catalyst.

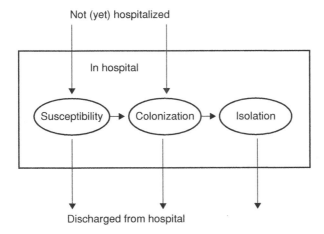

Figure 6.3 Schematic representation of the model developed by Forrester *et al.* (2007) in which individuals move between compartments representing their infection status, whether or not they are hospitalized and for those hospitalized whether or not they are in isolation following the detection of *Staphylococcus aureus*.

If $y(t)$ is the concentration of product at time t then

$$\frac{dy(t)}{dt} = \frac{v_{max} y(t)}{k + y(t)}. \tag{6.5}$$

Here v_{max} is the limiting reaction rate at high concentrations.

Depending on the nature of the data, the solution of equation (6.5) may be supplemented by the addition of a random term of mean 0 or sometimes by multiplication by a random term of mean 1. Depending again on the particular circumstances, the variance of the random term may depend on the explanatory variables and, especially when observations on the same individual are repeated over time, dependence between the random terms, especially at nearby time points, may need representation.

In other fields, in particular in the social sciences, processes tend to be so complicated that the formation of relatively simple equations to represent an underlying data-generating process is rarely possible. The study of the underlying data-generating processes is, however, a main route to understanding and proceeds empirically by building up dependencies step-wise in a way to be exemplified later. First, however, we give an atypical social science example where a relatively simple model is useful

Illustration: Social stochastic process   A situation in which a simple stochastic generating process is helpful in a social science context is a study of the size of conversational groups gathered at a social occasion (Coleman and James, 1961). The truncated Poisson distribution of the numbers of groups of size $1, 2, \ldots$ can be regarded as the equilibrium distribution of a particular stochastic process.

The following illustration is much more typical, however, and shows how a sequence of dependencies can be examined empirically.

Illustration: Hypothesized dependence structure   Figure 6.4 summarizes some aspects of a study done in the University of Mainz of the relationship between patient knowledge of diabetes and success in controlling the disease. Some of the variables recorded for each patient are named in largely self-explanatory form in the diagram. A key to the analysis is to determine for each possible pair of variables whether

- one of the two variables should be regarded as potentially explanatory to the other as a response, or
- the two variables should be treated as on an equal standing in the sense that, while they may be associated, the relationship is to be regarded as symmetrical.

The different boxes contain the main variables: first, there is a box containing intrinsic variables such as gender, and education; then there is a series of psychometric variables concerning attribution, how the individual perceives their responsibility concerning the disease; then there is a box containing a measure of knowledge of the disease; finally, there is an outcome, an objective measure of the glucose control achieved. The variables in any box depend in principle on all the variables in the preceding boxes, a so-called regression chain.

From this a recursive representation of the distribution of the full set of variables can be established in the form

$$f_{Y_q}(y_q) f_{Y_{q-1}|Y_q}(y_{q-1} \mid y_q) \cdots f_{Y_1|Y_2,\ldots,Y_q}(y_1 \mid y_2, \ldots, y_q). \qquad (6.6)$$

See Figure 6.4. Here Y_m is a collection of variables on an equal standing. The simplest case for discussion and interpretation is where each such

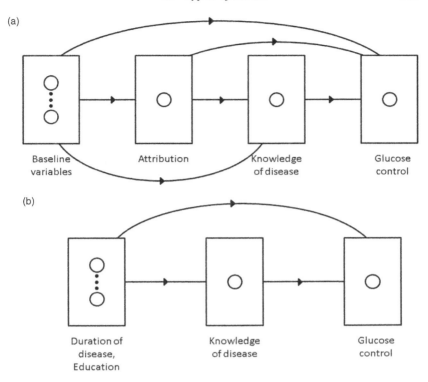

Figure 6.4 Relationship between diabetes control and explanatory features (based on Cox and Wermuth, 1996). As this was a cross-sectional study the direction of dependence is specified by hypothesis rather than by temporal ordering. (a) General formulation. The baseline variables were education, etc. The psychological attribution of disease was assessed by multiple test scores. Knowledge of the disease was assessed by a single test score. The variables (open circles) in a given box are of equal standing. Glucose control was an objective biochemical measure. (b) A simplified form after analysis. Glucose control depends on knowledge of the disease and on the baseline variables, that is, the duration of disease and the type of schooling, and on their interaction.

collection consists of a single component. Then each variable in the sequence is potentially dependent on all the variables preceding it. Quite often the initial variables Y_q may best be regarded as fixed, in which case the first factor in (6.6) is omitted. Provided each factor in (6.6) is specified by a separate set of parameters the model can be fitted recursively, that is, one factor at a time. In particular, if the components are single variables then fitting is by a sequence of univariate analyses.

We will return later to the limitations on the conclusions of this study, aspects of which illustrate the general difficulties of interpreting observational studies, especially those that are essentially cross-sectional, that is, involve measurement at a single time point for each individual.

In general regression chains such as those in the above illustration are intended to represent data-generating processes in the same spirit as differential equations do in the physical sciences.

A test of the meaningfulness of a possible model for a data-generating process is whether it can be used *directly* to simulate data. This test precludes, for example, models in which for two variables (X, Y) the value of Y appears in the equation defining X, together with random terms, and the value of X appears in the equation defining Y. Such a model may quite possibly define a joint probability distribution for (X, Y), but it is not possible to generate values, for example in a computer simulation, directly from this specification and therefore we ask: could a real system have worked in that way?

The above consideration precludes, in particular, certain types of simultaneous time series model. Therefore, to represent the feedback between two series (X_t, Y_t), a meaningful relationship should represent both components at time t as depending on both components at time $t - 1$, with a corresponding representation in differential equation form if the process is considered in continuous time.

6.4 Interpretation of probability

In the above discussion we have freely used the word *probability*. It is used in more than one way both in statistical contexts and in every-day life. One such meaning concerns the measure of uncertainty of knowledge about, for example, an event that may or may not occur or a hypothesis that may or may not be true.

In the formulation of models as a base for statistical analysis a different meaning is involved. Here, a probability is a proportional frequency of occurrence in repetitions of the process of observation, and, as we shall see, these repetitions may be quite closely realizable, may be somewhat hypothetical or may even be totally notional.

The essential point is that probability is used to represent, possibly in highly idealized form, a phenomenon in the real world. As such it is not essentially different from concepts like mass, force and energy or the density of a continuous medium, say a fluid, in which the limiting operation involved in defining the density at a point cannot be pushed to a molecular level.

Illustration: Repeated experimentation   In the botanical example concerning *Sordaria* (p. 98) the experiment in question can be repeated many times under the same conditions and the stability of the resulting frequencies can be demonstrated. The probability θ, defined as the limiting frequency in a process that could in principle be observed directly, is intended to capture an important property of the system.

In many fields the repetition involved in such an interpretation is notional rather than realizable in practice.

Illustration: Gradual replication   In the study of diabetic patients at the University of Mainz, repetition is feasible over time, as new patients accrue, and over different cities. In both cases, however, there may be systematic changes from the original study. The parameters of the probability model to be employed in the analysis aimed to capture the essential features of the situation as it was over the period studied, decoupled from the accidental features.

Illustration: Textual analysis   In a study of the works of Plato the distribution of the stress over the last five syllables of each sentence gave, for each work, a distribution over the 32 possibilities (stressed versus unstressed for each syllable) (Cox and Brandwood, 1959). Two large works, *Laws* and *Republic*, are known to have been written early and late respectively and a number of shorter works fall in unknown order in between. Let $\{p_{ir}\}$ denote the probabilities, assumed known; for *Laws* $i = 0$ and for *Republic* $i = 1$, for $r = 1, \ldots, 32$. Then for the kth smaller work suppose that the distribution is the exponential mixture

$$c(\theta_k) \exp\{\theta_k \log p_{1r} + (1 - \theta_k) \log p_{0r}\}, \tag{6.7}$$

where $c(\theta_k)$ is a normalizing constant. Note that at $\theta_k = 0, 1$ the distributions for *Laws* and *Republic* respectively are recovered. The values of θ_k place the works in order between the two large works, with the possibility of values outside $(0, 1)$ representing more extreme configurations. This leads to a simple and intuitively plausible index for placing the works in order relative to the two large works and could be regarded purely descriptively. Of course, interpretation of the order as being in fact temporal would raise further very large issues.

We are concerned here with the nature of the probability model. There is no sampling involved since the data are from the whole of the relevant works, nor are there are additional works by Plato. Clearly no explicit randomization is involved. The working hypothesis is that the process that generates the pattern of stresses is complicated and is such that it generates data *as if* the data were the consequence of a simple random process. To a limited extent this working hypothesis can be tested. If both the large works are divided into four sections then there results from each large work a 32×4 table showing the variation between sections within it. The standard chi-squared statistic from each such table has expectation equal to the degrees of freedom, 93; the calculated values are very close to this. That is, some features at least of the data-generating process are well represented by the probability model.

Rather similar considerations, namely the impossibility of effective repetition, apply for example in macroeconomic time series analysis where each, often quite short, section of each series is essentially unique. Believers in a particular economic theory might rely on that theory for support and even approximate verification of the model. A more empirical approach is to use the working assumption that at least some features of the series behave like the output of a random process and that the parameters of that process capture interesting features which are free of the accidents of the particular data being studied.

Illustration: Extrapolation of clinical trial results   In a simple form of clinical trial, patients giving informed consent are randomized between two treatments, say a new treatment T and a control C, and an outcome is recorded. Ignoring complications such as noncompliance with the allocated treatment or nonresponse, the data are analysed by a simple model for comparing two treatments that is appropriate for the kind of response involved, continuous, binary etc.

Suppose that reasonably strong evidence is obtained for the superiority, in the respect tested, of T over C. A conclusion that can be drawn with a very explicit statistical base concerns only the study individuals themselves. This is that had the whole group been given T the overall average outcome would have been better in some sense than had the whole group been given C. Any extension to other individuals would then be a qualitative judgement depending on, for example, an understanding of any underlying biological process; such a judgement would imply a general belief in reproducibility of conclusions.

Had the study individuals been chosen by some well-defined sampling procedure from a specific population of individuals, the conclusions would have a formal justification in being applied to that population. Note, however, that in a clinical trial, particularly because of the requirement of informed consent, the patients are unlikely to be close to a random sample of a target population even in the geographical area in question. Indeed in a formal sense they are unlikely to be chosen by any formal sampling procedure but, rather, to be individuals who happen to be available and willing to take part. In a somewhat similar sense, agricultural field trials are often done on experimental farms, which may be atypical of farming practice.

An intermediate position is that the observations are to be regarded as a sample from a *hypothetical* population of individuals and that the conclusions apply to that population. This expresses the notion that the conclusions do extend beyond the current individuals and are to some extent independent of the special nature of those individuals, even though the explicit character of the extended population cannot be specified.

In studies of a phenomenon at a relatively fundamental level it will often be best to start with situations in which the phenomenon may be most easily observed and to disregard considerations of the generalizability of conclusions.

Illustration: Generalizability of conclusions   Much early study of genetics used the fruit fly *Drosophila*. Results could be obtained quickly and relatively inexpensively, as contrasted with, say, work on mammals. While no doubt early workers considered to what extent their conclusions could be generalized to other species, no formal statistical issue of generalization arose.

The investigation concerning the culling of wildlife and bovine tuberculosis discussed above (see the illustrations on pp. 17, 39 and 77)
even though it had a strong policy-oriented motivation, chose as areas
for study those with the highest incidence. The choice was not made on
the basis of the representativeness of the cattle population. However, this
lack of representativeness was not of particular concern, as the areas of
greatest bovine tuberculosis incidence were of primary interest in terms
of targetting disease-control interventions.

It is an important principle of experimental design that if the wide applicability of conclusions is desirable then factors should be inserted into the
design that produce a range of experimental units and conditions enhancing
the range of validity of the conclusions.

6.5 Empirical models

6.5.1 Generalities

In many fields of study the models used as a basis for interpretation do not
have a special subject-matter base but, rather, represent broad patterns of
haphazard variation quite widely seen in at least approximate form. This is
typically combined with a specification of the systematic part of the variation, which is often, although not always, the primary focus of interest.
Giving the probabilistic part of the model then often reduces to a choice
of distributional form and of the independence structure of the random
components.

The functional form of probability distributions is sometimes critical, for
example when the prediction of extreme events is involved and in single-
parameter families of distributions, where there is an implicit assumption
of a relationship between variance and mean. This applies to the geometric,
Poisson and binomial discrete distributions and to the exponential continuous distribution. The assumption of any of these forms implies an indirect
estimate of variability without direct replication and, as such, is important
and possibly insecure. In many other cases the choice between alternative
parametric forms may be largely a matter of convenience.

To some extent, however, the simple situations that give rise to what
may be called the basic distributions of elementary probability theory, the
binomial, the Poisson, the geometric, the exponential, the normal and the
log normal, give some guide to empirical model choice in more complex
situations. Thus an initial analysis of the counts of haphazardly occurring

events in space or time may be based on the Poisson distribution modified for the systematic effects likely to be present.

In some specific contexts there is a tradition, possibly even supported by empirical success, establishing the form of model likely to be suitable. Sometimes the notion of *stylized facts* may be useful. These are broad generalizations which summarize the results of often complex statistical analyses without the details associated with caveats or special cases. Solow (1970) commented "There is no doubt that they are stylized, though it is possible to question whether they are facts."

Illustration: Time series For a financial time series of daily returns $Y(t) = \log\{P(t)/P(t - 1)\}$, where $P(t)$ is the price of a stock at time t, some stylized facts are as follows:

- the marginal distribution of $Y(t)$ is long-tailed;
- the serial correlations among the $\{Y(t)\}$ are small; and
- the serial correlations among the $\{Y^2(t)\}$ are appreciable.

A number of types of nonlinear time series correspond to these facts, although that is, of course, no guarantee of their suitability in any particular instance.

In other cases also, simple stochastic models may indicate a distributional form and the comparison of empirical data with theory may be intrinsically useful.

Illustration: Stochastic model suggests log normal distribution It can be shown that if systems of particles in, say, a soil are produced by a large number of proportional splittings of a system of large starting individuals, then the distribution of particle size will be log normal. Thus the comparison of an empirical distribution with the log normal form may be interesting in itself, as well as providing a useful base for further analysis.

6.5.2 Systematic variation

In many applications, although not in all, it is helpful to develop the random and the systematic parts of the model largely separately. Sometimes the systematic part, typically the aspects of the model that describe how response variables depend on explanatory variables, are strongly dictated by subject-matter theory. In many situations, however, it is a matter of setting out a sufficiently flexible empirical description, if at all possible containing parameters that have clear subject-matter interpretation.

The use of models with well-developed computer software can seem, in many contexts, to be almost obligatory and this tends to force analyses towards the use of standard procedures or minor adaptations thereof. The publically available R-project (R Development Core Team, 2007) makes available a wide range of traditional and newer methods, however.

Unless there is good reason otherwise, models should obey natural or known constraints even if these lie outside the range of the data. For example a regression relationship may be known to pass through the origin, even though the data, being remote from the origin, might suggest a relationship having, say, a positive intercept. In some cases, especially in the physical sciences where all or most variables are measured on a scale yielding positive values with a well-defined zero point, taking logs of such variables may be appropriate, yielding power-law-like relationships in which the coefficients, being dimensionless numbers, may have the advantage of clear interpretation. Such an argument would not justify taking logs of temperature in degrees Celsius even if these temperatures were all positive; in some contexts reciprocals of degrees Kelvin would be appropriate.

There can be conflicting considerations connected with such physical or logical constraints. For example, to describe the dependence of a binary response variable Y taking values 0 and 1 on a continuous explanatory variable x, the linear model

$$P(Y = 1) = \alpha + \beta x \tag{6.8}$$

incurs the constraint on its validity that for some values of x probabilities that are either negative or above 1 are implied.

For this reason (6.8) is commonly replaced by, for example, the linear logistic form

$$\log \frac{P(Y = 1)}{P(Y = 0)} = \alpha' + \beta' x, \tag{6.9}$$

which avoids any such constraint. The form (6.8) does, however, have the major advantage that the parameter β, which specifies the change in probability per unit change in x, is more directly understandable than the parameter β', interpreted as the change in log odds, the left-hand side of (6.9), per unit x. If the values of x of interest span probabilities that are restricted, say to the interval (0.2, 0.8), the two models give essentially identical conclusions and the use of (6.8) may be preferred.

If the relationship between a response variable and a single continuous variable x is involved then, given suitable data, the fitting of quite complicated equations may occasionally be needed and justified. For example, the growth curves of individuals passing through puberty may be quite complex in form.

In many common applications, especially but not only in the social sciences, the relationship between Y and several or indeed many variables $x_1, \ldots, x_p$ is involved. In this situation, possibly after transformation of some of or all the variables and preliminary graphical analysis, the following general ideas often apply:

- it is unlikely that a complex social system, say, can be treated as wholly linear in its behaviour;
- it is impracticable to study directly nonlinear systems of unknown form in many variables;
- therefore it is reasonable to begin by considering a model linear in all or nearly all of the $x_1, \ldots, x_p$;
- having completed the previous step, a search should be made for isolated nonlinearities in the form of curved relationships with individual variables x_i and interactions between pairs of variables.

The last step is feasible even with quite large numbers of explanatory variables.

Thus a key role is played in such applications by the fitting of linear relationships of which the simplest is the multiple linear regression, where for individual j the response Y_j is related to the explanatory variables $x_{j1}, \ldots, x_{jp}$ by

$$E(Y_j) = \beta_0 + \beta_1 x_{j1} + \cdots + \beta_p x_{jp}. \tag{6.10}$$

Here β_1, say, specifies the change in $E(Y)$ per unit change in x_1 with the remaining explanatory variables $x_2, \ldots, x_p$ held fixed. It is very important to appreciate the deficiencies of this notation; in general β_1 is influenced by which other variables are in the defining equation and indeed in a better, if cumbersome, notation, β_1 is replaced by $\beta_{y1|2\ldots p}$. If, say, x_2 were omitted from the equation then the resulting coefficient of x_1, now denoted by $\beta_{y1|3\ldots p}$, would include any effect on $E(Y)$ of the change in x_2 induced by the implied change in x_1. It is crucial, in any use of a relationship such as (6.10) to assess the effect of, say, x_1, that only the appropriate variables are included along with x_1.

For just two explanatory variables x_1 and x_2 the distinction between β_1 in the simpler equation and $\beta_{y1|2}$ is directly comparable with the relationship between the total and partial derivatives of functions of two (or more) variables. This is expressed by the equation

$$\beta_{y1} = \beta_{y1|2} + \beta_{y2|1}\beta_{21},\qquad(6.11)$$

where β_{21} is the formal regression coefficient of x_2 on x_1. This corresponds closely to

$$\frac{Dy(x_1,x_2)}{Dx_1} = \frac{\partial y(x_1,x_2)}{\partial x_1} + \frac{\partial y(x_1,x_2)}{\partial x_2}\frac{dx_2}{dx_1},\qquad(6.12)$$

the formula for total or directional differentiation of functions.

In some contexts different equations having similar forms may be under consideration to describe the same set of data; a simple special case is illustrated by (6.8) and (6.9). Sometimes examination of fit may indicate an unambiguous preference. In others the situation may be less clear. A general result is that if $\hat{\beta}_1, \hat{\beta}_2$ and $\hat{\beta}'_1, \hat{\beta}'_2$ denote the estimated parameters, all clearly nonzero, in two such models then

$$\hat{\beta}_1/\hat{\beta}_2 \simeq \hat{\beta}'_1/\hat{\beta}'_2.\qquad(6.13)$$

Essentially the reason is that both ratios measure the amount by which x_2 must change to induce the same change as a unit change in x_1 and this is not critically dependent on how the response is measured.

6.5.3 *Variational structure*

There are broadly three attitudes to random variation, the ultra-naïve, the naïve and the realistic. The first ignores such variation. This is sometimes appropriate, although probably rarely so in any context outlined in this book. The naïve approach assumes that apparently random variation does in fact correspond to truly random variation, thus in particular requiring one random variable per study individual, values for different individuals being assumed statistically independent. The often more realistic approach is to recognize the possibility of structure in the random variation. This may be a dependence between observations close together in time or space or a hierarchical structure corresponding to different levels of aggregation.

While there may be the great advantage of simplicity in the naïve approach and while such an approach is sometimes totally appropriate, particularly in carefully designed studies, on the whole ignoring structure in the random variation is likely to give misleading assessments of precision and in some cases may severely bias the conclusions.

A simple example illustrates the role of such dependence. The most basic formula underlying statistical analysis is that for n independent observations with the same mean and same standard deviation σ, the standard error of the mean is

$$\frac{\sigma}{\sqrt{n}}. \tag{6.14}$$

If, however, the observations are mutually correlated, this becomes

$$\frac{\sigma}{\sqrt{n}}(1 + \Sigma\rho_{ij})^{1/2} \tag{6.15}$$

where ρ_{ij} is the correlation coefficient between observations i and j and the sum is over all possibilities with $i \neq j$, so that each pair counts twice. Thus if, for example, each observation is correlated with k other observations with correlation coefficient ρ then (6.15) gives a standard error of approximately

$$\frac{\sigma}{\sqrt{n}}(1 + k\rho)^{1/2}. \tag{6.16}$$

Thus, especially if k is appreciable, correlation can induce a major change, which with positive correlation is an increase, in the error of estimation. Similar results apply to estimated regression coefficients and to more complicated correlation patterns. The effect on, for example, confidence intervals for parameters can be appreciable, typically leading to underestimation of variability. Negative correlation is less likely in most contexts and produces a decrease in the variance of the mean. Thus in some sampling situations, and in particular in simulation studies, design to introduce negatively correlated estimates by the use of so-called *antithetic variables* is advantageous.

Illustration: Negative correlation   In an animal feeding study in which animals are caged in groups and food is provided separately for each cage there may be negative correlations between the weight gains of two animals in the same cage, arising from competition for limited resources.

The general implications of this are partly the desirability of steps in design to avoid correlated errors, where these are indeed consequences of the measurement process, and partly the need for a reasonably realistic representation of patterns of variability where these are natural and part of the system under study.

6.5.4 Unit of analysis

In many applications it is important to be explicit about the unit of analysis. This, in particular, affects the level of definition appropriate for the response and other variables and the level of modelling detail; it also has a bearing on the important independence assumptions involved in model formulation. Different definitions of the unit of investigation may be needed at different stages of analysis, depending on the research questions involved.

In a randomized experiment the unit of investigation is the smallest subdivision of the material such that any two units *might* receive different treatments. In some types of observational study the unit of investigation may be defined by considering what would have been done in a corresponding randomized experiment had one been possible.

Illustration: Individual and cluster randomization   In a randomized clinical trial, if individual patients are randomized to one of two or more treatments then the patient is the unit of analysis. If all patients in the same clinic receive the same treatment, through a system known as cluster randomization, the clinic is the unit. In a cross-over design in which patients receive one treatment for a period and a possibly different treatment in a second period the unit is a patient–period combination.

Illustration: Multiple units of analysis in the same experiment   In a study of the possible effects of the culling of wildlife on the incidence of bovine tuberculosis, 30 roughly circular areas were chosen in sets of three, each being approximately 100 km^2 in area and containing about 100 farms. Within each set of three areas one was randomized to proactive culling, another to reactive culling (that is culling only in response to a tuberculosis occurrence in cattle) and one was a no-culling control. The unit of analysis for the randomized comparison was the circular area. In particular the primary variable for analysis was the number of TB cases in the whole area.

In fact some data were obtained also on positions within the areas of potentially infected wildlife. To study the distribution of TB incidence *within* the circular areas involves non-randomized comparisons and the use of a farm as the unit of analysis rather than an aggregate of farms, as in the circular area.

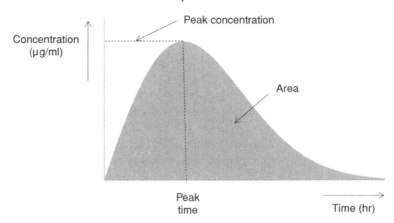

Figure 6.5 Blood concentration following injection. The data can be characterized by fitting an equation or by summarizing statistics such as peak concentration, time at which the peak was achieved and total area under the concentration–time curve.

In some contexts there may be a clear hierarchy which establishes the level of unit appropriate for each research question.

Illustration: Hierarchical data structure   Suppose that data are obtained on certain characteristics of school children, the children being arranged in classes within schools with an appreciable number of schools within each of several districts (Koukounari *et al.*, 2006). For different purposes the unit of analysis may be the child, the class or the school. However, if the number of districts were very small, comparisons between them would be a largely descriptive issue.

Two general issues are that the assessment of precision comes primarily from comparisons between units of analysis and that the modelling of variation within units is necessary only if such internal variation is of intrinsic interest.

Illustration: Summarizing response curves   In a comparison of the impact of a number of different related compounds, suppose that the study units are rats. For each rat, one compound is injected and then the concentration in the blood of a critical substance is measured hourly for the next 24 hours; see Figure 6.5.

To compare the treatments the complex response for each rat must be reduced to a simpler form. One way that does not involve a specific model is to replace each set of values by, say, the peak concentration recorded, the time at which that peak is recorded and the total area under the concentration–time curve. Comparison of the treatments would then hopefully be accomplished, for example by showing the differences in some and the constancy of others of these distinct features. An alternative approach would be to model the variation of concentration with time either by an empirical equation or perhaps using the underlying differential equations. The treatment comparison would then be based on estimates of relevant parameters in the relationship thus fitted.

The general issue involved here arises when relatively complex responses are collected on each study individual. The simplest and often the most secure way of condensing these responses is through a number of summary descriptive measures. The alternative is by formal modelling of the response patterns. Of course, this may be of considerable intrinsic interest but inessential in addressing the initially specified research question.

In other situations it may be necessary to represent explicitly the different hierarchies of variation.

Illustration: Hierarchical determinants of the response variable   An investigation of the factors affecting the judicial sentencing of offenders (Johnson, 2006) allowed for the hierarchical nesting of sentences within judges as well as the nesting of judges within courts or counties. While this study found that individual-level variables, for example prior criminality, had the strongest effects on sentencing, for understanding the observed variation the judge- and county-level variables were important.

Illustration: Hazards of ignoring hierarchical data structure   An analysis of the success of in vitro fertilization pre-embryo transfer (IVF-ET) used hierarchical logistic regression models to analyse the effect of fallopian tube blockage, specifically hydrosalpinx, on the probability of embryo implantation success. Variation was modelled both at the individual-woman level and the embryo level, allowing for variability in women's unobservable probability of implantation success. Hogan and Blazar

(2000) found evidence of considerable heterogeneity among women, which if ignored would lead to biased parameter estimates and underestimates of their associated standard errors.

Notes

The role of empirical statistical models is discussed in many books on statistical methods. For a general discussion of different types of model with more examples, see Lehmann (1990) and Cox (1990).

7

Model choice

Detailed consideration is required to ensure that the most appropriate parameters of interest are chosen for a particular research question. It is also important to ensure the appropriate treatment of *nonspecific effects*, which correspond to systematic differences that are not of direct concern. Thus in the present chapter we discuss aspects relating to the choice of models for a particular application, first the choice between distinct model families and then the choice of a specific model within the selected family.

7.1 Criteria for parameters

7.1.1 Preliminaries

In some applications analysis and interpretation may be based on nonparametric formulations, for example the use of smooth curves or surfaces summarizing complex dependencies not easily captured in a simple formula. The reporting of estimated spectral densities of time series or line spectra representing complex mixtures of molecules are examples. Mostly, however, we aim to summarize the aspects of interest by parameters, preferably small in number and formally defined as properties of the probability model. In the cases on which we concentrate, the distribution specified by the model is determined by a finite number of such unknown parameters.

For a specific research question, parameters may be classified as *parameters of interest*, that is, directly addressing the questions of concern, or as *nuisance parameters* necessary to complete the statistical specification. Often the variation studied is a mixture of systematic and haphazard components, with attention focused on the former. In such a case the parameters of interest will concern the systematic variation. The nuisance parameters relate to the haphazard variation together with any aspects of the systematic variation not of direct interest. The roles are reversed, however, when attention is focused on the haphazard variation.

The choice of parameters involves, especially for the parameters of interest, their interpretability and their statistical and computational properties. Ideally, simple and efficient methods should be available for estimation and for the assessment of the precision of estimation.

7.1.2 Parameters of interest

Commonly, the parameters of interest assess dependencies, for example distinctions in the response patterns for different levels of an explanatory variable or, in other cases, rates of change in the response. It is essential that the subject-matter interpretation of a parameter is clear and that it is measured in appropriate units, which should always be stated, and it is preferable that the units are chosen to give numerical answers that are neither inconveniently small nor inconveniently large. If the zero point of a scale of measurement has a special meaning, this must be respected. For example, the transformation of temperatures from degrees Kelvin to degrees Celsius is not always appropriate.

For response variables that are *extensive*, that is physically additive, the mean value will typically be an aspect of its distribution that is of interest.

Illustration: Interpreting analyses of extensive variables Whatever the shape of the distribution of the yield of a product per plot in an agricultural field trial, yield is an extensive variable; that is, the yield of the combined area of two plots is the sum of the separate yields. This implies that, say, the differences between the yields experienced over a large area from two different treatments is estimated by the difference in the means per unit area times the total area. There may be a clash here with other interpretative aspects in that, for example, analysis might show that over replication of the comparison in different contexts it is the ratio of the yields that is stable, not the difference. That is, it might be that one treatment gives a particular fractional increase in yield rather than a particular difference. Then, even if the final interpretation is in terms of differences, care is needed regarding the generalizability of the conclusions.

Illustration: Expressing impact using the population attributable fraction Assessment of the effect of exposure to specific risk factors is often specified in terms of a ratio or percentage effect, but the public health interpretation for a specific population in terms of the numbers of individuals affected is in terms of a difference in rates, not a ratio. A preference for ratios, simple dimensionless numbers, such as the factor 10 increase in the risk of lung cancer due to cigarette smoking, stems partly from a greater ease of intuitive understanding, partly from the potential relative stability of the ratio across different populations and partly from a desire to assess the possible effect of unobserved confounders. Nevertheless the public health importance of such an effect in terms of the number of individuals affected per, say, 10^3 individuals at risk requires a *difference* not a *ratio* of probabilities. To address this need, epidemiologists often calculate the population attributable fraction, that is, the proportion of the disease risk that would be eliminated from a population if the risk factor(s) under study were eliminated (Greenland and Robins, 1988).

In some contexts, especially with quantitative variables with a natural origin, it may be possible to achieve representations of systematic variation in which the parameters are dimensionless and indeed are simple integers or simple rational numbers. This is achieved by fitting relationships of the form

$$y \propto x_1^{\beta_1} \cdots x_p^{\beta_p}. \tag{7.1}$$

Here the indices are intrinsically dimensionless, that is, unchanged by a change in units of any component variable.

The logarithmic transformation of all variables then leads to a linear relationship, often with simple additive error structure. A logarithmic transformation is suggested by the assumed systematic structure. If the transformation happened to produce normality of the distribution of departures and constancy of the conditional variance of y, that would be a bonus.

When, all variables are measured in units derived from the basic units of mass, length and time, as would be the case in some physical and biological problems, the principles of dimensional analysis will imply constraints among the coefficients in (7.1).

Illustration: Constraints on parameters arising from physical dimensions
The flow of a fluid along a tube may be smooth (laminar) or erratic (turbulent). The type of flow depends among other things on the velocity of flow x_1, the radius of the tube x_2, the density x_3 and the viscosity

x_4 of the fluid. If to a first approximation the probability that the flow is turbulent is a function of

$$x_1^{\beta_1} \cdots x_4^{\beta_4} \tag{7.2}$$

then the dimensionless character of any probability function imposes constraints on the β_i.

To see this we will use $[x]$ to denote the dimensions of the variable x; in a physical context this will be some combination of length, time and mass. In fact, viscosity is a force per unit area and force has the dimensions of mass times acceleration, i.e. $[M][L][T^{-2}]$. That is,

$$[x_1] = [L][T]^{-1}, \qquad [x_2] = [L],$$
$$[x_3] = [M][L]^{-3}, \qquad [x_4] = [M][L]^{-1}[T]^{-1}.$$

Thus the dimensions of (7.2) are

$$[L]^{\beta_1 + \beta_2 - 3\beta_3 - \beta_4}[M]^{\beta_3 + \beta_4}[T]^{-\beta_1 - \beta_4}$$

and the dimensionless requirement leads to

$$\beta_1 = \beta_2 = \beta_3 = -\beta_4,$$

and thus to consideration of the quantity:

velocity times radius times density divided by viscosity.

This quantity is called the Reynolds number. Empirical experience is that Reynolds numbers in the high hundreds usually correspond to turbulent flow. The dimensional argument is not commonly presented in a statistical context but in fact applies to any simple model of the systematic component of variability of measurements on standard physical types of scale.

We will illustrate some of these points, in particular the importance of considering the units of measurement, first by a simple linear regression.

Illustration: Interpreting the slope parameter of a linear regression In the straight-line model (6.2),

$$E(Y_j) = \beta_0 + \beta_1 x_j, \tag{7.3}$$

the parameter β_1 represents the change in expected response corresponding to a unit change in the explanatory variable x. It is measured in units of y per unit of x and, when the units are derived from the standard units of mass, length and time, it is conventional to write the dimensional form as $[\beta_1] = [y][x]^{-1}$. An appropriate choice of specific units, for example kilograms, grams etc., may be largely conventional in a specific field but should, if possible, be chosen to ensure that β_1, if this quantity is of primary concern, is not far outside the range $(1/10, 10)$.

The general meaning of β_1 is as set out above. The more precise meaning depends strongly on the specific subject-matter context and in particular on the study design. The strongest interpretation, in a reasonable sense causal, would be justified if the study individuals had been independently randomized to levels of x. Then β_1 would quantify the difference between the hypothetical responses of the same individual under two levels of x that are a unit apart. In other contexts the interpretation would be much weaker, being confined either to empirical prediction of y over future individuals drawn from the same population or to assessing the effect of changing x while allowing for unspecified variables related both to x and to y to change appropriately.

As an example of dimensional considerations consider a simple situation involving survival or response times.

Illustration: Interpreting parameters from analyses of failure time data   In many situations, as noted above, times from a defined origin to a clearly specified outcome, for example system failure or patient death, have a long-tailed distribution. An initial model for such observations may then be the exponential distribution with density $\rho e^{-\rho y}$ and mean $1/\rho$. Note that $[\rho] = [T^{-1}]$, so that ρ is measured, say, as a rate per year. This form will apply if failure occurs in a Poisson process, that is, as the first of a completely random series of point events. If the distribution is long-tailed but not exponential then the next step may be to consider a two-parameter family. There are many possibilities, in particular:

- a gamma distribution;
- a log normal distribution; or
- a Weibull distribution

All have their uses and it may be difficult to choose between them empirically. The log normal distribution does not include the exponential distribution as a special case and has for some, but not all, purposes the disadvantage of leading to an analysis that is very sensitive to small survival times.

We use the Weibull family to illustrate some points about parameterization. The family is defined by the survivor function

$$P(Y > y) = \exp\{-(\rho y)^{\gamma}\} \tag{7.4}$$

and hence by the density

$$\gamma(\rho y)^{\gamma-1} \exp\{-(\rho y)^{\gamma}\}. \tag{7.5}$$

In this expression γ is dimensionless, that is, it is unchanged by a change in the units of time, say from hours to days. It can be regarded as defining the shape and dispersion of the distribution relative to that of the exponential distribution, the special case $\gamma = 1$. Because ρy is also dimensionless, $[\rho] = [T^{-1}]$ is a rate per unit time.

Of course, many other parameterizations are possible. For example the survival function could be written as

$$\exp(-\alpha y^{\gamma}).$$

While this is perfectly adequate as a base for numerical fitting, dimensionality arguments lead to $[\alpha] = [T^{-1/\gamma}]$ showing that substantive interpretation of the numerical value of α will be difficult, if not impossible.

One desirable property of a parameter of interest is that it can be estimated with reasonable precision from specific sets of data and that its value remains relatively stable across distinct situations.

Especially in many situations involving cross-classified explanatory variables, symmetry may suggest redundant parameterizations, those having more parameters than are estimable. Numerical fitting usually depends on the use of free parameters and hence the imposition of constraints that break the symmetry of the original form. An additional feature is that the revised form may lead to estimates with a relatively complicated error structure involving, as a minimum, a covariance matrix of errors and not just a variance for each estimated parameter.

Illustration: Equivalent parameterizations   A simple example is provided by a normally distributed response variable y that is dependent on two explanatory variables, both of which take a discrete set of levels. In a symmetrical form of model with no interaction, if Y_{kl} represents an observation at levels k, l respectively of the two variables then

$$E(Y_{kl}) = \mu + \alpha_k + \beta_l. \tag{7.6}$$

Potentially, the extension of such a model to more than two factors is useful much more broadly.

Arbitrary constants may be added separately to α_k and β_l and subtracted from μ, leaving the model unchanged. Note that our interest is focused on internal contrasts among the two sets of parameters and that these contrasts are unaffected by the reparameterization.

In the absence of special considerations, a common approach is to choose baseline levels of the two explanatory variables, say (1, 1); note that in unbalanced data the chosen levels should have appreciable frequency of occurrence in the data. The model then can be written as

$$E(Y_{kl}) = \nu + \alpha_k^* + \beta_l^*, \tag{7.7}$$

where

$$\alpha_1^* = \beta_1^* = 0.$$

The new parameters can now be estimated together with their standard errors. This is all that is needed if in fact the only interest is in comparing the levels with the baseline. If other contrasts are to be examined, for example level 3 with level 2, a point estimate is directly obtained as, for example, the difference in estimators $\hat{\alpha}_3 - \hat{\alpha}_2$ but the standard error requires an estimate of the covariance $\text{cov}(\hat{\alpha}_3, \hat{\alpha}_2)$. While in principle there may be no difficulty in obtaining the latter, it is not commonly specified, other than optionally, in standard software; in reporting conclusions it may be impracticable to record full covariance matrices of estimates, especially in large systems.

Often the difficulty can be largely avoided by a different, so-called floating, reparameterization in which for estimating, say, contrasts among the α_k we write

$$E(Y_{kl}) = \alpha_k^{**} + \beta_l^*, \tag{7.8}$$

where the α_k^{**} are unconstrained. The $\hat{\alpha}_k^{**}$ may then have very small covariances.

7.2 Nonspecific effects

7.2.1 Preliminaries

A common issue in the specification of models concerns aspects of the system under study that may well correspond to systematic differences in the variables being studied but which are of no, or limited, direct concern.

Illustration: Effects arising from variables of little or no direct interest
A clinical trial may involve several or many centres, an agriculture field trial may be repeated at a number of different farms and over a number of growing seasons and a sociological study of the relationship between class and educational achievement may be repeated in broadly similar form in a number of countries. In a laboratory study different sets of analytical apparatus, imperfectly calibrated, may be used. Here the centres, farms, seasons, countries and sets of apparatus may well correspond to important differences in outcome but are not the direct object of study.

One general term for such features is that they are *nonspecific*. Two different centres in a trial may vary in patient mix, procedures commonly used, management style and so on. Even if clear differences in outcome are present between clinics, specific detailed interpretation will be at best hazardous.

It may be necessary to take account of such features in one of two different ways. The simpler is that, on an appropriate scale, there is a parameter representing a shift in level of outcome. The second and more challenging possibility is that the primary contrasts of concern, treatments, say, themselves vary across centres; that is, there is a treatment–centre interaction. We will deal first with the simpler case.

The possibility arises of representing the effects of such nonspecific variables by random variables rather than by fixed unknown parameters. In a simple example, within each of a fairly large number of centres, individuals are studied with a binary response and a number of explanatory variables, including one representing the treatment of primary concern. For individuals in centre m the binary response Y_{mi} might be assumed to have a logistic regression with

$$\log\left\{\frac{P(Y_{mi} = 1)}{P(Y_{mi} = 0)}\right\} = \alpha_m + \beta^T x_{mi}, \tag{7.9}$$

where x_{mi} is a vector of explanatory variables, the vector of parameters β is assumed to be the same for all centres and the constant α_m characterizes the centre effect. A key question concerns the circumstances under which the α_m should be treated as unknown constants, that is, *fixed effects*, and those in which it is constructive to treat the α_m as random variables representing what are therefore called *random effects*.

7.2.2 Stable treatment effect

In a fairly general formulation when there is assumed to be no treatment–centre interaction, the outcome of interest is represented by a combination, usually linear, of a treatment effect, a centre effect and terms representing other features of concern. This might be a direct representation, for continuous approximately normally distributed variables, or a linear logistic representation, for binary variables.

We are not concerned here with the technical details of how such models are fitted but rather with whether parameters representing centre effects should be regarded as 'fixed' parameters on the same conceptual standing as other parameters or as random variables. Typically the latter involves treating the centre effects as independent and identically distributed random variables, often normally distributed with unknown mean and variance although of course other possibilities are available.

Effective use of such a random-effects representation will require estimation of the variance component corresponding to the centre effects. Even under the most favourable conditions the precision achieved in that estimate will be at best that from estimating a single variance from a sample of a size equal to the number of centres. This suggests that use of a random-effects representation will be very fragile unless there are at least, say, 10 centres and preferably considerably more.

Next, if centres are chosen by an effectively random procedure from a large population of candidates, individual distinctions between which are of little interest, then the random-effects representation has an attractive tangible interpretation. This would not apply, for example, to the countries of the European Union in a social survey. The countries would be of individual interest and the total number of such countries small. Even if the countries sampled showed essentially the same effect of interest, extrapolation to *all* countries of the EU would require judgement about possible anomalous behaviour of the countries omitted from the study.

It is important to note that in relatively balanced situations the distinction between treating centres as random or fixed often has little or no effect on the formal conclusions about the treatment effect. This can be seen in its simplest form in the traditional randomized block design shown in Table 7.1(a). There are $t = 3$ alternative treatments and the experimental units are arranged in $b = 5$ blocks each consisting of $t = 3$ similar units. A key feature is that each treatment occurs once in each block.

For continuous variables, a least-squares analysis of a model in which each observation is the sum of a treatment effect and a block effect and

Table 7.1 *(a) Randomized block design with t = 3 treatments in b = 5 blocks. (b) An unbalanced arrangement with a similar structure*

(a)	Block	1	2	3	4	5
		T_2	T_1	T_3	T_3	T_1
		T_1	T_2	T_1	T_2	T_3
		T_3	T_3	T_2	T_1	T_2
(b)	Block	1	2	3	4	5
		T_1	T_2	T_1	T_2	T_3
		T_1	T_3	T_2	T_2	T_2
		T_2	T_3	T_3	T_3	T_1

also has a random error leads to estimation of the parameters via respectively the treatment means and the block means. It is fairly clear on general grounds, and can be confirmed by formal analysis, that estimation of the treatment effects from the corresponding marginal means is unaffected by whether the block effects are:

- arbitrary unknown constants;
- random variables with an arbitrary distribution; or
- some known and some unknown constants, for which all or some block differences are unimportant.

For the first two possibilities, but not the third, estimation of the residual variance is also unaffected. Thus the only difference between treating the block constants as random rather than fixed lies in a notional generalization of the conclusions from the specific blocks used to a usually hypothetical population of blocks.

If, however, the complete separation of block and treatment effects induced by the balance of Table 7.1(a) fails, the situation is different. See Table 7.1(b) for a simple special case with an unrealistically small number of blocks! The situation is unrealistic as the outcome of an experiment but is representative of the kinds of imbalance inherent in much comparable observational data.

If now the block parameters are treated as independent and identically distributed with variance, say, σ_B^2 and the residuals in the previous model have variance σ_W^2 then all observations have variance $\sigma_B^2 + \sigma_W^2$ and two observations have covariance σ_B^2 if and only if they are in the same block and are otherwise uncorrelated. The so-called method of generalized least squares may now be used to estimate the parameters using empirical estimates of the two variances. Unless σ_B^2/σ_W^2 is large the resulting treatment effects are not those given by the fixed-effect analysis but exploit the fact that the block parameters are implicitly assumed not to be greatly different from one another, so that some of the apparent empirical variation between the block means is explicable as evidence about treatment effects. It is assumed that reasonable estimates of the two variances are available. If in fact the ratio of the variances is very small, so that inter-block variation is negligible, then, as might be expected, the block structure can be ignored and the treatment effects again estimated from marginal means.

Indeed in the older experimental design literature the above procedure, presented slightly differently, was called the *recovery of inter-block information*.

We now move away from the relatively simple situations typified by a randomized block design to consider the broad implications for so-called mixed-model analyses and representations. The supposition throughout is that in the model there are explanatory variables of direct and indirect interest and also one or more nonspecific variables that are needed but are typically not of intrinsic interest. The issue is whether such variables should be represented by unknown parameters regarded as unknown constants or by random variables of relatively simple structure, typically, for each source independent and identically distributed. The following considerations are relevant.

- Unless there is appropriate replication, estimation of the variance of the random effect terms cannot be effectively achieved. Such estimation is required implicitly or explicitly for an analysis based on the random effects model to be secure.
- If two analyses, one treating the nonspecific effects as fixed but nonzero and the other assuming them all to be zero, give essentially the same estimates of the effect of important explanatory variables and the relevant standard errors, then it is unlikely that a random effects model will give anything substantially different.
- In other cases, however, it is in principle possible that a random effects analysis will give either appreciably different estimated effects or

improved estimates of precision and, moreover, estimated effects mostly intermediate between the two fixed effect analyses.

- Representation of the nonspecific effects as random involves independence assumptions which certainly need consideration and may need some empirical check. If in applications of the type represented in Table 7.1 the blocks are all the same size and treatments are allocated in accordance with a randomized design in which there are more treatments than units per block, then this ensures that the block effects may be treated as random. However, in an observational study such randomization is not available. If, for example, in an observational study the blocks contained differing numbers of units and the block effects were larger in blocks with more units then bias would be introduced by assuming the block effects to be totally random. That apart, the crude analysis suggested in the second list item will often give a pointer to the potential gains of the random effects formulation.

With widely available software all these procedures are relatively easily implemented unless the data are either extremely extensive or of very complex form. The point of the above discussion is that it is important in applications to understand the circumstances under which different methods give similar or different conclusions. In particular, if a more elaborate method gives an apparent improvement in precision, what are the assumptions on which that improvement is based? Are they reasonable?

7.2.3 Unstable effect

The previous discussion presupposed that the effects of the explanatory variables under study are essentially the same across varying levels of the nonspecific variable. If this is not the case, that is, if there is an interaction between an explanatory variable and a nonspecific variable, the first step should be to explain this interaction, for example by transforming the scale on which the response variable is measured or by introducing a new explanatory variable characteristic of the levels of the nonspecific variable. That is, a deterministic explanation should be sought.

Illustration: Interpreting a variable treatment effect   Two medical treatments compared at a number of centres show appreciable discrepancies between centres in the treatment effect, measured as a ratio of the two event rates. Is an explanation, for example, that it is the *difference* of event rates that is stable rather than the ratio? Or does the ratio depend in a systematic way on the socio-economic make-up of the patient population at each centre? While any such explanation is typically tentative it will usually be preferable to the other main possibility, that of treating interaction effects as random.

An important special application of random-effect models for interactions is in connection with overviews, that is, the assembling of information from different studies of essentially the same effect. Many issues discussed in this book are in fact relevant to overviews, but we shall not develop the topic in detail.

7.3 Choice of a specific model

Having considered the nature and properties of probability models, we now turn to the choice of models for fitting to a specific application. Often this will involve at least two levels of choice, first between distinct separate families and then between specific models within a chosen family. Of course all choices are to some extent provisional. As we have seen, in many situations the systematic and random components of variability require distinct, even if related, choices.

Illustration: Choosing between model families   For the analysis of survival data that are approximately exponentially distributed, both the gamma family and the Weibull family provide distributions with two parameters to be estimated. The distributions are separate except in the special case of exponentially distributed data. As an example, for the representation of the dependence of a continuous response Y on an explanatory variable x, the two families

$$E(Y) = \beta_0 + \beta_1 x, \qquad E(Y) = \gamma_0/(1 + \gamma_1 x) \qquad (7.10)$$

are separate unless there is no dependence on x, that is, $\beta_1 = \gamma_1 = 0$. Formal tests to determine which is the more suitable family of models are probably rarely necessary. When considering two families of models, it is important to consider the possibilities that both families are adequate, that one is adequate and not the other and that neither family fits the data. Informal comparison of the maximized log likelihoods achieved under the two models may be helpful, especially if the numbers of adjustable parameters in the two models are the same. For a formal test of a model family specified by β against an alternative family specified by γ the best procedure is often to simulate sets of data from the first model using the estimated values $\hat{\beta}$ and to compare the maximum log likelihoods achieved with the second family with those obtained from the empirical data. The procedure should then be repeated, interchanging the roles of the two models. Such comparisons are sometimes made using Bayes factors, which aim to give the probability of correctness of each model. Bayes factors are, however, misleading if neither model is adequate.

We have discussed previously the use of specific subject-matter considerations to determine a suitable model. Even in more empirical situations there may be special circumstances, appeal to which may be valuable.

For dependencies of Y on x that are gently curved, or for testing the adequacy of a linear representation, a low-degree polynomial, in particular a quadratic, may be entirely adequate. But if it is known on general grounds that the relationship approaches an asymptote for large x then the form

$$E(Y) = \alpha + \gamma e^{-\delta x} \tag{7.11}$$

may lead to a more stable interpretation.

In a different situation it may be known that any general relationship must pass through the origin, yet a straight line not through the origin may give an entirely adequate fit, if for example all the observed values of x are positive and quite far from zero. If all observations are positive, a linear relationship of $\log Y$ on $\log x$ may be preferred because it automatically satisfies the external constraint even if there is no improvement in fit by its use.

We concentrate now on the more widely occurring issue of how to make a choice within a specified family using formal techniques.

In some contexts there is a family of models with a natural hierarchy of increasing complexity, and the specific issue is to decide how far down the hierarchy to go.

Illustration: Polynomial regression models Systematic polynomial relationships in one or more explanatory variables provide a flexible family of smooth relationships for empirical fitting, although it should be noted that polynomials often provide a particularly poor base for even modest extrapolation. Thus with one explanatory variable, x, measured from some suitable reference point, we may fit

$$E(Y) = \beta_0 + \beta_1 x + \cdots + \beta_p x^p \qquad (7.12)$$

to the data.

With two explanatory variables, x_1 and x_2, a comparable relationship in the case $p = 2$ is

$$E(Y) = \beta_{00} + \beta_{10} x_1 + \beta_{01} x_2 + \beta_{20} x_1^2 + 2\beta_{11} x_1 x_2 + \beta_{02} x_2^2.$$

It will typically be wise to measure the x_i from a meaningful origin near the centre of the data. The β_{jk} then specify first- and second-order partial derivatives of the fitted function at that origin.

Somewhat similarly, time series may be represented by autoregressive processes of order $p = 1, 2, \ldots$ or by more complicated mixed autoregressive-moving average processes.

In all these expressions, unless there are special considerations, all terms of degree up to and including the final term will be used. For example, in the second equation it would not normally be sensible to include β_{11} and to exclude β_{20} and β_{02}. For both equations we have a clear hierarchy of models corresponding to increasing values of p.

For a single set of data it would be typical to use in such a model the smallest value of p achieving adequate consistency with the data, and there are a number of rather different procedures for assessing this. An exception to the general strategy might be where, say, a linear representation is adequate but it is required to put limits on the amount of curvature that might be present.

Often, however, there are a number of similar sets of data, each of which may be represented in the above form. Typically it is then desirable to fit all these sets by a model with the same value of p, usually close to the largest value needed in any individual set. The practice of, for example, fitting straight lines, $p = 1$, to some sets and quadratics, $p = 2$, to others, and so on, is potentially confusing and may establish false distinctions between different sets as well as making comparison of estimates of specific parameters difficult.

In the case of the dependence of a response on qualitative features, the hierarchical principle implies, again with very rare exceptions, that models with interaction terms should include also the corresponding main effects.

Illustration: Implication of no main effect in the presence of two-way inter-action terms In an experiment in which animals are fed different diets, where for each diet there are equal numbers of males and females, and gain in body weight is measured for each animal, the outcomes could be summarized in a two-way table of means. In this the rows would be defined by diet and the two columns by gender. A model with in-teraction but no diet (row) effect would represent the possibility that, although the differences between diets are not the same for males as for females, nevertheless averaged over males and females the diets give *exactly* the same mean gain, an implausible possibility. If a limited amount of each food were fed collectively to single groups of ani-mals then the non-hierarchical form would be, perhaps, slightly more plausible.

Illustration: Requirement for no row effect despite the inclusion of column and interaction terms Suppose that in an industrial setting a rectangu-lar piece of material is cut into horizontal strips, the cuts being exactly the same distance apart. Suppose that it is also divided by vertical lines that are nominally the same distance apart but which in fact vary errati-cally. Thereby nominally equal rectangular small pieces of material are formed. Suppose that these are weighed and the weights compiled in a two-way table of rows and columns. Then the procedure as described would require a representation with zero row effects but nonzero column and interaction terms.

In the discussion that follows we assume that, where applicable, a hierar-chical principle is satisfied by all models in that, for example, interactions in a model are accompanied by the corresponding main effects.

We now discuss in more detail the choice of model within a family of re-gression models of a response y on a set of explanatory variables $x_1, \ldots, x_p$, where p may be quite large. The same principles apply to least-squares and other forms of generalized linear models, those dealing for example with binary data or survival data. We will assume that data become available on n individuals. In most cases n is much greater than p; we will comment later on the other possibility.

Suppose that, at some point in the analysis, interest is focused on the role of a particular explanatory variable or variables, x^* say, on the response, y. Then the following points are relevant.

- The value, the standard error and most importantly the interpretation of the regression coefficient of y on x^* in general depends on which other explanatory variables enter the fitted model.
- The explanatory variables prior to x^* in a data-generating process should be included in the model unless either they are conditionally independent of y given x^* and other variables in the model or are conditionally independent of x^* given those other variables. These independencies are to be consistent with the data and preferably are plausible *a priori*.
- The variables intermediate between x^* and y are to be omitted in an initial assessment of the effect of x^* on y but may be helpful in a later stage of interpretation in studying pathways of dependence between x^* and y.
- Relatively mechanical methods of choosing which explanatory variables to use in a regression equation may be helpful in preliminary exploration, especially if p is quite large, but are insecure as a basis for a final interpretation.
- Explanatory variables not of direct interest but known to have a substantial effect should be included; they serve also as positive controls. Occasionally it is helpful to include explanatory variables that should have no effect; these serve as negative controls, that is, if they do not have the anticipated positive or null impact then this warns of possible misspecification.
- It may be essential to recognize that several different models are essentially equally effective.
- If there are several potential explanatory variables on an equal footing, interpretation is particularly difficult in observational contexts.

It is helpful in relatively complicated cases to distinguish at least two phases in the analysis. The first phase consists of the search among perhaps several possibilities for a base for interpretation. In the second phase the adequacy of that base is checked.

For the first phase a number of fairly different routes may reach the same goal. One broad approach is as follows. Suppose that there are explanatory variables x^* which we require to include in a model, because of their intrinsic interest, and other variables $\tilde{x}$ which are potential confounders, that is, they are conceptually prior to x^* and may influence both x^* and y.

Here is one strategy set out in a series of steps:

- fit a reduced model $\mathcal{M}_{\text{red}}$ with only x^*;
- fit, if possible, a full model $\mathcal{M}_{\text{full}}$ with x^* and $\tilde{x}$;

- compare the standard errors of the estimated regression coefficients of y on x^* in the two models.

Except possibly for minor consequences connected with the estimation of a residual mean square, the standard errors from $\mathcal{M}_{\text{red}}$ will be less than those from $\mathcal{M}_{\text{full}}$, and possibly much less. If in fact there is relatively little change in standard error between the two models then the use of $\mathcal{M}_{\text{full}}$ may seem the more secure base for interpretation. It may be helpful in showing conclusions to give adjusted and unadjusted estimates of important parameters alongside one another.

If, however, the reduced model gives substantial improvements in notional precision, as is likely to be the case if p is large, it is worth exploring the potential omission of some components of $\tilde{x}$. A mixture of the backwards elimination of components of $\tilde{x}$ and the reintroduction of components by forward selection, starting from none or a few of the components of $\tilde{x}$ may be used. The ultimate object is, we repeat, assessment of the effect of the variables, x^*, rather than the study of the components of $\tilde{x}$.

A more challenging situation arises when potentially a large number of components of x^* are of concern.

Illustration: Analysis of an extensive case-control questionnaire   To study the possible effect of farm management practices on the incidence of bovine tuberculosis, farmers having experience of a recent outbreak were compared with farmers who had not had an outbreak by the use of a comprehensive questionnaire about the nature of their farm and its management. Retrospective case-control studies of this kind are analysed by logistic regression of the formal binary response, outbreak versus no outbreak, on the explanatory variables, the questionnaire answers, themselves mostly taken in binary form. The questionnaire was extensive and simplification by omission of some components unavoidable. To implement the broad strategy set out above, the variables could be divided into x^*, those that represent management features that are in principle modifiable, and $\tilde{x}$, those that are in a sense intrinsic to the farm, the nature of the soil, the size and location of the farm and so on.

When the dimension of x^* is greater than one then, even though choosing an appropriate model may raise no special difficulties, interpretation of the estimated parameters may be ambiguous, especially in observational studies.

In some studies, especially in the process industries, specification of the effect of several quantitative factors on the response may best be understood by considering a response surface specifying the expected response as a function, linear or often quadratic in the components of x^*. The coefficients of individual components of x^* refer to the change in the response consequent on changes of one component variable with the others held fixed, for example the change consequent on a change in temperature of a chemical process with the pressure and concentrations of the reactants held fixed. It may be better if the emphasis in these studies is on the surface itself rather than on the individual coefficients and indeed it may be helpful to transform x^* into a new set of derived variables that will clarify description of the surface.

The situation is formally similar in observational studies but typically the interpretation of the parameters is more difficult.

Illustration: Characterizing a response surface   Consider the relationship between blood pressure y and the sodium, Na, and potassium, K, levels in the blood, denoted jointly by x^*. From suitable data a response surface may be fitted, possibly linear on a log scale of all variables, or perhaps more complicated, and possibly adjusted for background variables. Interpretation of this surface and its associated parameters depends greatly on the context. We now illustrate some key principles in idealized form.

Suppose first that in an experimental setting the values of Na and K are in intakes per day and that for each study individual these values are controlled at a set level for each individual, the levels being randomized to individuals according to a suitable experimental design. The regime is continued until the systolic blood pressure has come to an equilibrium; this value is taken as the response variable of interest. Then the regression parameters in the fitted model have a direct interpretation as representing the mean change in blood pressure consequent on a specified change in Na and K intake, for example on a unit increase in Na with K fixed.

Suppose next that the observations are on a collection of study individuals whose daily intake of Na and K is measured over a period. Their blood pressure is also recorded. This is an observational setting; the daily intakes depend on the diets chosen by the individuals in question and are not assigned by the investigator. The initial method of statistical analysis of such data might be the same as for the experiment described above. The interpretation of the parameters is, however, more restricted.

For example, any demographic or lifestyle variable considered to be prior to Na and K in a data-generating process can in principle be included in the model if measured and hence in effect held fixed when notional changes in Na and K are considered. Any such variables affecting both Na, K intake and blood pressure and which are unobserved distort the conclusions. Thus if the amount of exercise taken is not recorded and influences both Na and K levels and blood pressure, a notional change of, say, Na used to interpret a regression coefficient includes an implied but unobserved change in the amount of exercise.

If the individual data values refer to distinct individuals, the data would estimate the response surface connecting blood pressure with the Na and K concentrations. If the analysis were on a log scale then the extraction of meaningful combinations of the two concentrations would be possible. In general, however, interpretation of the individual regression coefficients, assuming that both are appreciable, would be difficult without further information, because the notion of changing, say, Na, while holding K fixed is unlikely to correspond to a real situation.

A particular issue that may arise at the initial stage of an analysis concerns situations in which a number of conceptually closely related measures are recorded about the same aspect of the study individuals.

Illustration: Characterizing a multi-faceted risk factor   Smoking behaviour can be recorded in a number of different ways. Examples of relevant variables include: current smoker, yes or no; for previous smokers, years since cessation; cigarettes, including number smoked, pipe, and/or cigar; inhaler, yes or no. It would not be sensible to treat these as entirely separate variables. If the effect of smoking is a key issue under study, careful attempts to isolate which of the variables, or combinations thereof, provides the best explanation are required. If, however, data on smoking is included only to avoid the contamination of other effects, some initial simplification to one or two summary measures is desirable. In either case it would be unwise to treat the variables as conceptually unrelated, even though correlated.

For the second phase of analysis at least two different considerations arise. First certain explanatory variables $\tilde{x}$ that were potential candidates for inclusion in the model may have been omitted. It will be wise to add them back into the model, one (or a very small number) at a time, and to check whether any major change in the conclusions is indicated. Second, and very important, the initial discussion was predicated on a linear fit. Now, especially in relatively complex situations it is unlikely that linearity of response is other than a crude first approximation. While specific nonlinear effects can, of course, be included from the start, a general form of nonlinearity with many explanatory variables is not a feasible base for analysis, unless p is small. A compromise is that, probably at the end of the first phase of analysis but perhaps earlier, nonlinear and interaction effects, such as are expressed by x_j^2 and $x_j x_k$, are explored one at a time by their addition to the model as formal explanatory variables to describe curvature of the response to the explanatory variables. The resulting test statistics, assessed graphically, provide a warning process rather than a direct interpretation; for example, the detection of interaction signals the need for detailed interpretation.

The choice of a regression model is sometimes presented as a search for a model with as few explanatory variables as reasonably necessary to give an adequate empirical fit. That is, explanatory variables that are in some sense unnecessary are to be excluded, regardless of interpretation. This approach, which we do not consider, or in general recommend, may sometimes be appropriate for developing simple empirical prediction equations, although even then the important aspect of the stability of the prediction equation is not directly addressed.

A topic that has attracted much interest recently is that of regression-like studies in which the number, p, of explanatory variables exceeds the number, n, of independent observations. In these applications typically both n and p are large. Some key points can be seen in the simple case studied in the industrial experimental design literature under the name 'supersaturated design'. In this one might wish to study, for example, 16 two-level factors with only 12 distinct experimental runs feasible. If it can be safely assumed that only one of the 16 factors has an important effect on response then, especially if a suitable design (Booth and Cox, 1962) is chosen, there is a good chance that this factor can be identified and its effect estimated. If, however, several factors have appreciable effects then it is likely that confusing and potentially misleading results will be obtained. That is, success depends crucially on the *sparsity* of the effects. Whether it is ever wise in practice to use such a design is another matter!

It has been shown recently that for general linear regression problems with $p > n$ and with sparse effects, that is, most regression coefficients are negligible, estimation and identification of the nonzero components is possible but the sensitivity of the answers to moderate failure of the sparsity requirement is often unclear; see, however, Meinshausen and Bühlmann (2010).

Notes

Section 7.1. For more discussion of the choice of parameters, see Ross (1990). Dimensional analysis is described in textbooks on classical physics. For a general discussion of the use of floating parameters, see Firth and de Menezes (2004). There is an extensive literature on the analysis of survival data, variously with engineering, medical or social science emphasis. Kalbfleisch and Prentice (2002) gave a thorough account in a largely medical context. For the comparison of different parametric forms, see Cox and Oakes (1984).

Section 7.2. For recovery of inter-block information, see Cox and Reid (2000).

8

Techniques of formal inference

Assessment of the uncertainty implicit in conclusions is often an important feature of statistical analyses. Although the primary emphasis in this book is not on the specialized concepts and methods involved in such assessments, the present chapter reviews the main ideas involved under the following headings: confidence limits, posterior distributions and significance tests.

8.1 Preliminaries

The details of specific statistical methods and the associated mathematical theory will not be discussed here. We shall, however, outline the main forms in which statistical conclusions are presented, because understanding of the strengths and limitations of these forms is essential if misunderstanding is to be avoided.

We discuss first analyses in which interpretation centres on individual parameters of interest; that is, in general we investigate component by component. We denote a single such parameter by ψ. This can represent some property of interest, such as the number of individual animals of a specific wildlife species in a particular area, or it can represent contrasts between groups of individuals in an outcome of interest or, in a linear regression application, ψ can be the slope of the relationship.

It is desirable that ψ is considered in 'sensible' units, chosen to give answers within or not too far outside the range from 0.1 to 10. For example, a slope of a relationship of length against time, that is a speed, might be in mm per hour or km per day, etc., as appropriate and an incidence rate might be in the number of cases per 100 000 person-years or in cases per 100 person-days. It is good practice always to state and repeat units explicitly.

Then conclusions about ψ can be presented as:

- confidence limits or intervals at one or more levels, for example 95%, 99%, etc.;

- a summary, formally similar to the above, of a posterior distribution for ψ;
- a significance test of a null hypothesis, that ψ takes a specific value ψ_0 of interest; for example this could be $\psi_0 = 0$, corresponding to zero slope of a regression line;
- an estimate t and its estimated standard error s_t.

We will discuss these in turn. The last is often, but not always, able in effect to encompass the others; the third, the significance test, is by far the most likely to be misunderstood. In the final section of the chapter we discuss the complications that can arise affecting interpretation of these procedures. We use idealized numerical examples as illustrations.

8.2 Confidence limits

The most satisfactory form of specification using confidence limits is to give an upper limit and a lower limit at conventionally chosen levels, thus forming an interval. For example, an upper 97.5% limit and a lower 2.5% limit combine to form a 95% confidence interval. In some cases only the upper limit, say, is of interest. Unfortunately, in many cases only the level of the interval is given, without any explanation of how the upper and lower limits separately have been specified, but this is not ideal. There is, in principle, no restriction to using a single confidence level and in most cases a *confidence distribution* can, in principle at least, be considered.

The empirical interpretation of, for example, a 97.5% upper limit is that it is calculated by a procedure which in a long run of repeated applications would give too small a value in only about 2.5% of cases. As such it summarizes information about ψ provided by the data *considered on their own*, always, of course, in the light of the model assumptions involved. The issues involved in examining sets of confidence intervals for related parameters derived from different studies will be discussed later.

Consideration, at least in principle, of a confidence distribution shows that in most cases the values near the centre of a confidence interval are more likely than those at the extremes and that if the true value is actually above the reported upper limit it is not likely to be far above it. Conventional accounts of confidence intervals emphasize that they are not statements of probability. However, the evidential impact of, say, a 97.5% upper confidence limit is much the same as that of an analogous upper probability, or credible, limit. The crucial distinction is that different confidence-limit statements, even if statistically independent, may not be combined by the laws of probability.

The following complications arise occasionally and need special discussion.

- It can happen that values of ψ which are reasonably consistent with the data fall into two or more disjoint intervals, in which case specification simply by a single upper limit and a single lower limit is unsatisfactory.
- In some situations no allowable parameter values are consistent with the data.
- In some situations, notably in the estimation of a ratio as in some bio-assays and in the method of instrumental variables in econometrics, the natural confidence region may sometimes consist of all values *except* those in a specified interval. Further, in some cases the data are such that no value should be excluded from the set of values reported as consistent with the data. These superficially anomalous conclusions are in fact a natural consequence whenever the denominator of a ratio is badly determined.

Illustration: Comparing counts with the Poisson distribution A rather idealized version of a counting problem in particle physics occurs when the number of events counted in a particular time has a Poisson distribution with a mean that is the sum of a background contribution of known mean and a signal, which may be zero. If the observed count is substantially *less* than the expected background then a reasonable conclusion may be that the data are inconsistent with the proposed model. The consequence would be the need to check the proposed background assessment and possible complications in the recording process.

Confidence limits for functions of multiple parameters such as the ratio α/β should never be approximated by plugging the univariate confidence limits into the functional form. Instead local linearization, the delta method, can be used to obtain asymptotic means and variances for nonlinear functions of random variables.

8.3 Posterior distributions

The posterior distribution of a parameter is in most contexts the parallel to a confidence distribution, forming what used to be called an inverse probability statement. This is nowadays called the Bayesian approach. In it the model specification is augmented by a probability distribution, termed the prior distribution, representing information about the parameters other than that supplied by the data under analysis. The availability of this distribution enables the standard rules of probability theory to be applied to determine the probability distribution as revised in the light of the data, termed the posterior distribution. There are a number of different possibilities. Sometimes the prior distribution is 'flat' over a particular scale, representing the absence of initial information. Then in relatively simple problems the posterior distribution is virtually equivalent to the confidence distribution, both numerically and in interpretation. In complex problems with many nuisance parameters the use of flat prior distributions is suspect and, at the very least, needs careful study using sensitivity analyses.

The use of historical data to form a prior distribution is equivalent to a pooled analysis without, however, the necessary check of mutual consistency; if the prior distribution and the data are in conflict then absurd conclusions are likely to arise. The insertion of expert opinion directly, in the form of a probability distribution rather than through the data on which that expertise is based, raises interesting possibilities especially in investigations with a strong decision-making focus. A key requirement is that there should be sufficient agreement among experts that elicitation of their opinions merits consideration on an equal footing with the empirical data. For interesting examples, see Choy *et al.* (2009) and Martel *et al.* (2009).

The general issue is not whether any external information should be used in interpretation. Rather it is whether that information can and should be reasonably represented by a probability distribution and the resulting information seamlessly integrated with that from the data.

A mild note of warning is perhaps needed over the use of the adjective *Bayesian* in current statistical discussions. Thomas Bayes (1701–1761) related the conditional probability of an event A given an event B to the conditional probability of B given A and the overall probabilities of the individual events. He then applied this formula to a special inference problem to obtain the conditional probability of an explanation given the data from the probability of the data given the explanation, i.e. from the implied model and from the individual probabilities, in particular here the probability of the explanation without knowledge of the data, that is, the prior probability of the explanation.

In the nineteenth century this method, applied to problems of statistical inference, was called the method of inverse probability and, as noted above, used a prior distribution intended to represent initial ignorance. This latter notion was severely criticized by nineteenth-century mathematicians such as Boole and Venn and, while never abandoned totally, fell into disfavour in most of the first half of the twentieth century. The most important exception was the work of Jeffreys (1939), who attempted a systematic justification of an approach in which probability represents an objective degree of belief. In the middle of the century attention shifted to a personalistic view of probability, as concerned with *individual* belief as expressed by individual choice in a decision-making context. It was at this point that the adjective *Bayesian* came into common use, replacing the term 'inverse probability'.

Towards the end of the twentieth century the personalistic view received much less emphasis, probably because of its inappropriateness for scientific discussion whatever its virtues might be for individual decision making or even, perhaps, for describing what an individual scientist in his or her heart of hearts really believes, a different matter from what is reasonably soundly based on evidence. The word *Bayesian*, however, became ever more widely used, sometimes representing a regression to the older usage of 'flat' prior distributions supposedly representing initial ignorance, sometimes meaning models in which the parameters of interest are regarded as random variables and occasionally meaning little more than that the laws of probability are somewhere invoked.

When a statistical discussion or analysis is described as Bayesian, the following points should be checked in order to avoid any misunderstanding of the meaning of the procedure.

- Is the prior distribution a positive insertion of evidence? If so, what is its basis and has the consistency of that evidence with the current data been checked?
- If so-called indifference or ignorance or reference prior distributions have been used, how have they been chosen? Has there been a sensitivity analysis? If the number of parameters over which a prior distribution is defined is appreciable then the choice of a flat prior distribution is particularly suspect and indeed potentially misleading.
- Each of a substantial number of individuals may have been allocated a value of an unknown parameter, the values having a stable frequency distribution across individuals. The use of a so-called empirical Bayesian argument to estimate the parameters for specific individuals, or contrasts of such values, is clear cut here.
- In some contexts the Bayesian approach is probably best seen as a computationally convenient way of obtaining confidence intervals.

8.4 Significance tests

8.4.1 Types of null hypothesis

Suppose that there is a value ψ_0 of the parameter ψ of special interest. We call the hypothesis $\psi = \psi_0$ the *null hypothesis*, H_0. To understand significance tests it is essential to distinguish between different types of null hypothesis and, slightly less importantly, to deal separately with the different levels of problem formulation.

In one context, common in some applications, H_0 divides the parameter space into two qualitatively different regions. These may, for example, be regions of positive slope as contrasted with regions of negative slope or regions in which treatment A gives a higher (perhaps clinically more beneficial) mean outcome than treatment B versus the contrary. There may be no special reason for thinking the null hypothesis to be even approximately true. Its importance stems from the implication that, so long as H_0 is reasonably consistent with the data, the sign or direction of the effect under study has not been securely established. In most cases use of a significance test is essentially equivalent to recording the level at which the confidence interval just overlaps zero, that is, the data are reasonably consistent with effects in both possible directions. In one sense, therefore, no special discussion is required. In Bayesian discussions the analogue of a null hypothesis involves the posterior probability that the true value is in one particular part of the set of possible values. As noted previously this will often yield conclusions nearly identical to those from non-Bayesian discussion.

In the second type of null hypothesis there is distinct interest in the possibility that H_0 is exactly true or can be treated as true to an adequate approximation. For example, H_0 may correspond to the prediction of a firmly based theory. Another possibility is that there may be strong reasons for supposing that a certain modification to a system has no effect on a specified response variable. We shall call any such null hypothesis *atomic*.

Illustration: Testing an atomic null hypothesis   Maconochie and Roman (1997) analysed data on singleton births in Scotland for the period 1975 to 1988. The null hypothesis was that gender occurred completely randomly at a fixed probability per birth. They tested for dependence on the genders of previous siblings in the family, on birth order, on maternal age and social class and on year and period of birth. From data on over half a million births they found no evidence against the null hypothesis.

8.4.2 *Test of atomic null hypothesis*

In the narrowest interpretation of an atomic null hypothesis, only that null hypothesis is formalized; for example it could represent the consequences of one theory, no alternative theory being available or even at the stage of formal development. Typically, some idea of the departures of interest will be available, but very possibly only qualitatively.

It is worth drawing a parallel between the testing of an atomic null hypothesis and the testing of a comparable deterministic hypothesis. For the latter, the procedure would be as follows:

- find a property, v, whose value, v_0, is predicted by the theory and which is expected to be, say, larger than v_0 if the theory is false;
- measure v;
- if $v = v_0$, the data are consistent with the theory but if $v > v_0$ then the hypothesis is to be rejected and the theory that generated it is itself either rejected or modified; or
- if unexpectedly $v < v_0$, reconsideration of the whole approach may be appropriate.

If $v = v_0$ then the data could be regarded as not merely consistent with the theory but as providing support for the theory to the extent that no alternative explanations are available, that is, that any alternative explanation would have been detected by the design of the study.

In the case of a statistical atomic null hypothesis H_0 we may aim to parallel the above steps. For this, the procedure is as follows:

- find a function v of the data such that under H_0 the corresponding random variable V has, to an adequate approximation, a known distribution, for example the standard normal distribution, or, in complicated cases, a distribution that can be found numerically;
- collect the data and calculate v;
- if the value of v is in the lower or central part of the distribution, the data are consistent with H_0; or
- if v is in the extreme upper tail of the distribution then this is evidence against H_0.

The last step is made more specific by defining the p-value of the data y, yielding the value v_{obs}, as

$$p = p(y) = P(V \geq v_{\text{obs}}; H_0). \tag{8.1}$$

It is assumed that this can be found numerically, at least to a reasonable approximation. The *p*-value is in a sense the unique measure of the extremity of the value of v in the light of H_0.

More explicitly, the *p*-value has the following hypothetical interpretation.

If we were to accept the current data as barely decisive evidence against H_0 then out of a large number of cases in which H_0 is true, it would be falsely rejected in a long-run proportion p of cases.

In this formulation the statistic v is chosen on largely intuitive grounds as one that is likely to be sensitive to the kinds of departure from H_0 of interest. If there is a parametric model in which a parameter ψ represents the focus of interest then the statistic used to form the best confidence intervals for ψ will be used for v, and the *p*-value essentially specifies the level at which a confidence interval for ψ just fails to contain the null value ψ_0.

Some accounts of the philosophy of science put an emphasis on the testing and especially the rejection of atomic null hypotheses. This emphasis may stem from a preoccupation with the physical sciences, with their powerful theory base. Inconsistency with the null hypothesis may then point constructively to minor or not so minor modification of a theoretical analysis. In many other fields atomic null hypotheses are less common, and significance tests are more often used to establish the directions of effects and may best be considered as subsidiary to confidence distributions.

8.4.3 Application and interpretation of p-values

The outcome of a significance test should be reported by giving the *p*-value approximately. The conventional guidelines are roughly as follows:

- if $p \simeq 0.1$ there is a suggestion of evidence against H_0;
- if $p \simeq 0.05$ there is modest evidence against H_0;
- if $p \simeq 0.01$ there is strong evidence against H_0.

The following points of interpretation arise.

- The values mentioned above are not to be interpreted rigidly. The interpretation of 0.049 is not essentially different from that for 0.051.
- This use of tests to assess evidence must be distinguished from a decision-making rule, as for example in industrial inspection, in which inevitably somewhat arbitrary choices have to be made at the borderline. There is, however, no justification in that kind of application for standard

choices such as the level 0.05. The critical level that is sensible depends strongly on context and often on convention.

- The p-value, although based on probability calculations, is in no sense the probability that H_0 is false. Calculating that would require an extended notion of probability and a full specification of the prior probability distribution, not only of the null hypothesis but also of the distribution of unknown parameters in both null and alternative specifications.

- The p-value assesses the data, in fact the function v, via a comparison with that anticipated if H_0 were true. If in two different situations the test of a relevant null hypothesis gives approximately the same p-value, it does not follow that the overall strengths of the evidence in favour of the relevant H_0 are the same in the two cases. In one case many plausible alternative explanations may be consistent with the data, in the other very few.

- With large amounts of informative data, a departure from H_0 may be highly significant statistically yet the departure is too small to be of subject-matter interest. More commonly, with relatively small amounts of data the significance test may show reasonable consistency with H_0 while at the same time it is possible that important departures are present. In most cases, the consideration of confidence intervals is desirable whenever a fuller model is available. For dividing null hypotheses, the finding of a relatively large value of p warns that the direction of departure from H_0 is not firmly established by the data under analysis.

8.4.4 Simulation-based procedures

The calculation of p-values and associated quantities such as confidence intervals is based on mathematical theory. The procedures may be *exact* or may derive from so-called *asymptotic* or *large-sample* theory. Here 'exact' means only that no mathematical approximations are involved in the calculation. Of course, any calculation is made in terms of a model, which at best can be only a very good approximation, so that its exactness must not be overinterpreted. Large-sample procedures involve mathematical approximations, essentially that the effect of random variation on estimates and test statistics is relatively small, so that nonlinear functions are locally linear, and that some associated quantities are normally distributed. It is probably rare that these approximations have a critical effect on interpretation; specialized advice is needed when, for example, the level of significance achieved or the width of confidence intervals are judged to be critical.

An alternative approach uses computer simulation instead of mathematical theory. Such procedures are of a number of types. The simplest simply mimics the mathematical theory. That is, to test a null hypothesis, for example, repeat observations corresponding in structure to the data and for which H_0 is satisfied are made again and again, the test statistic is calculated for each and its distribution is built up until the position of the data in the simulated distribution is sufficiently clear.

An appealing approach due to Efron (1979), which uses only the data under analysis, that is, not drawn afresh from the probability model, goes under the general name of *bootstrap*. The reference is probably to the old expression 'pulling yourself up by your own shoe-laces'. Here, repeat hypothetical data are constructed by sampling the real data repeatedly with replacement to generate, for example, the distribution of an estimate or test statistic. In sufficiently simple situations this gives a close approximation to the results of mathematical analysis. It does, however, not cope so easily with situations having a complex error structure; whether an answer achieved in this way is more secure than one obtained by mathematical analysis is by no means always clear.

In Bayesian approaches broadly similar issues arise. A posterior distribution that cannot be obtained in an mathematical exact form may be approximated, typically by the method called Laplace expansion, or simulated by the powerful computer-intensive technique of Markov chain Monte Carlo (MCMC). This leads to the simulation of pseudo-data from which the posterior distribution of the quantities of interest can be built up. There may be issues about whether the convergence of certain processes has been achieved, and the whole procedure is a black box in that typically it is unclear which features of the data are driving the conclusions.

8.4.5 Tests of model adequacy

Formal statistical analysis rests, as we have repeatedly emphasized, on the formulation of a suitable probability model for the data. The question whether a model is adequate or whether it should be modified or indeed even replaced by a totally different model must always be considered. In some situations informal inspection may be adequate, especially if the conclusions from the analysis are relatively clear cut. The key issue is then whether there could be an aspect of the model which, if changed, would alter otherwise clear-cut conclusions.

If a formal test is needed, the model is taken as the null hypothesis and the principles outlined above are followed. Occasionally two quite different model families will be available. Then the most fruitful approach is usually to take each model as the null hypothesis in turn, leading to a conclusion that both are consistent with the data, or that one but not the other is consistent or that neither is consistent with the data. Another possibility is to form a single model encompassing the two. Methods are sometimes used that compare in the Bayesian setting the probabilities of 'correctness' of two or more models but they may require, especially if the different models have very different numbers of parameters, assumptions about the prior distributions involved that are suspect. Moreover such methods do not cover the possibility that all the models are seriously inadequate.

More commonly, however, there is initially just one model family for assessment. A common approach in regression-like problems is the graphical or numerical study of residuals, that is, the differences between the observed and the fitted response variables, unexpected patterns among which may be suggestive of the need for model modification. An alternative, and often more productive, approach is the use of test statistics suggestive of explicit changes. For example, linear regression on a variable x can be tested by adding a term in x^2 and taking the zero value of its coefficient as a null hypothesis. If evidence against linearity is found then an alternative nonlinear representation is formulated, though it need not necessarily be quadratic in form.

The status of a null hypothesis model as dividing or atomic depends on the context. Some models are reasonably well established, perhaps by previous experience in the field or perhaps by some theoretical argument. Such a model may be regarded as an atomic null hypothesis. In other situations there may be few arguments other than simplicity for the model chosen, and then the main issue is whether the direction in which the model might be modified is reasonably firmly established.

8.4.6 Tests of model simplification

A type of application in a sense complementary to that in the previous subsection is that of model simplification. We shall describe this initially in the context of multiple regression, although the considerations involved are quite general. Suppose that we consider the linear regression of a response variable Y on explanatory variables $x_1, \ldots, x_d$. A residual sum of squares, $RSS_{1\ldots d}$, will result. Suppose now that a reduced model of interest uses as explanatory variables a subset of the original variables, say $x'_1, \ldots, x'_{d'}$, where $d' < d$. This will have a larger residual sum of squares, $RSS_{1'\ldots d'}$. We call the difference

$$\text{RSS}_{1'...d'} - \text{RSS}_{1...d} \tag{8.2}$$

the sum of squares for the included variables *adjusted for* the omitted variables, that is, those not in $(x'_1, \ldots, x'_{d'})$. If this sum of squares is too large, there is evidence that the improvement in fit from the more complex model is too large to be a chance effect and thus the proposed simplification is in conflict with the data. A formal test uses the variance-ratio, or F, distribution to compare (8.2) with the residual sum of squares from the full model, $\text{RSS}_{1...d}$, allowing for the numbers of parameters (degrees of freedom) involved.

In a much more general setting we have a full model and a reduced model and we obtain the maximized log likelihoods under each, l_{full} and l_{red}, say. Then under the null hypothesis that the reduced model has generated the data, the test statistic

$$2(l_{\text{full}} - l_{\text{red}}) \tag{8.3}$$

has an approximately chi-squared distribution with the number of degrees of freedom equal to the number of parameters suppressed in the simpler model, that is $d - d'$. This result requires some technical conditions which may need checking in applications.

Occasionally Occam's razor is put forward as a philosophical justification for the relentless simplification of models. In some situations where it is thought that some relatively simple process is involved, drastic simplification to the essentials may indeed be wise. In many applications, however, for example in observational studies in the social sciences, the statistical models are already very considerably idealized models of complex phenomena and the wisdom of further simplification is less clear. There are two rather different reasons which may even in such situations demand simplification.

One concerns presentation. It is distracting to list a large number of regression coefficients most of which are small and not interpretable. This argument applies only when d is not small. If when a model is simplified the subject-matter interpretation remains essentially the same as for the full model, that fact should be reported even if the details are omitted.

The second reason is that if there is appreciable internal correlation in the explanatory variables then a substantial nominal gain in the precision with which key coefficients are estimated may be achieved by excluding apparently irrelevant variables from the model. This is best assessed by comparing, for parameters important for interpretation, the estimates and standard errors obtained from the fit of a full model with those obtained from one or more simplified models. If it is essential to follow this approach, so far as is feasible insensitivity to the particular choice of simplification should be examined.

The approach to mathematical formulation taken in virtually all statistical work may be contrasted with that in classical applied mathematics, that is, mathematical physics in which relativistic and quantum effects are negligible. There, the key equations are Newton's laws of motion, the Navier–Stokes equations of fluid dynamics and Maxwell's equations of electrodynamics. Thus if the forces acting on a moving body are measured, together with its mass and acceleration, and if force is not equal to mass times acceleration then this is taken to imply either a missing component or a misspecification of force. This defect must if possible be defined and identified. That is, a lack of fit implies a deficiency in the data, not a defect in Newton's laws. This approach has, of course, been outstandingly successful. By contrast, in the contexts that we are considering a clash between the data and the model nearly always represents a defect in the model rather than in the data.

8.5 Large numbers of significance tests

8.5.1 Generalities

There are a number of situations which, at first sight at least, amount to examining the outcomes of a large number of significance tests. In all such cases it is desirable to see whether a formulation in terms of estimation may not be preferable.

In a microarray study of a particular disease, data from diseased and control subjects are compared at a large number of loci. Is there evidence of any association and if so at which loci? In a genome-wide association study (GWAS) (Wellcome Trust Case Control Consortium, 2007) somewhat similar statistical issues arose with a very large number of loci.

Illustration: Testing multiple energy cells   Data arising from particle colliders, such as the large hadron collider (LHC), require complex preprocessing but can be considered as Poisson-distributed counts corresponding to events collected into a large number of energy bands. There is a background count rate, which is a smooth function of energy that, to a first approximation, may be assumed known. Interest lies in the energy cells in which the observed counts exceed the background. A very large number of individual cells may need to be examined unless the energy associated with the events of specific interest is known.

Illustration: Testing multiple plant varieties A plant breeding programme may start with field trials of a very large number of varieties. On the basis of the yields (and other properties) a subset of varieties is chosen for a second phase, and so on, until one variety or a relatively small number of varieties are chosen for possible commercial use.

In all these examples, except probably the last, the issue may appear to be how to interpret the results of a large number of significance tests. It is, however, important to be clear about objectives. In the last illustration, in particular, a decision process is involved in which only the properties of the varieties finally selected are of concern.

8.5.2 Formulation

In fact, despite the apparent similarity of the situations illustrated above, it is helpful to distinguish between them. In each situation there is a possibly large number r of notional null hypotheses, some or none of which may be false; the general supposition is that there are at most a small number of interesting effects present. Within that framework we can distinguish the following possibilities.

- It may be that no real effects are present, that is that all the null hypotheses are simultaneously true.
- It may be virtually certain that a small number of the null hypotheses are false and so one must specify a set containing false null hypotheses, such a procedure to have specified statistical properties.
- The previous situation may hold except that one requires to attach individual assessments of uncertainty to the selected cases.
- The whole procedure is defined in terms of a series of stages and only the properties of the overall procedure are of concern.

8.5.3 Multi-stage formulation

Many applications are essentially multi-stage, in that provisional conclusions from the first stage of data collection and analysis in principle need confirmation.

Illustration: Lack of independent replication of findings   Webb and Houlston (2009), in a review of statistical considerations in modern genetical research connected with cancer, report, disturbingly, that relatively few reported conclusions about the relationship between single-nucleotide polymorphisms (SNPs) of candidate genes and cancer have been confirmed in independent replication.

If, however, as in the plant-breeding illustration mentioned above, the whole procedure is explicitly planned on a multi-stage basis to be implemented in a relatively predetermined way, the relevant statistical properties of the whole procedure should be assessed; issues of statistical significance at each stage will not be directly relevant.

Illustration: Clinical trial phases   The development of pharmaceutical products through a series of phase 1, phase 2, phase 3 clinical trials might, for some general managerial purposes, be usefully modelled as a multi-stage procedure. In any specific instance, however, the decision to move from one phase to the next would involve careful assessment of the evidence available at that particular phase.

8.5.4 Bonferroni correction

The following discussion applies to situations in which a global null hypothesis may be effectively true. The argument can be put in its simplest form as follows.

Suppose that independent predictors are available. Each is tested for statistical significance in r independent tests and the most significant effect found is at level p^*, that is, p^* is the smallest of the individual levels achieved. Suppose, for example, $r = 20$ and that p^* is approximately 0.02 for a particular predictor variable x^*.

Now, had x^* been tested on its own then fairly clear evidence against the relevant null hypothesis would have been found. Equivalently, a new investigator approaching the topic with only x^* in mind as a potential explanatory variable would reach the same conclusion. Suppose, however, that x^* is considered solely because it corresponds to the smallest p-value. Then it is clear that if the hypothetical interpretation of p-values used to motivate their use, namely their relationship to the probability of false rejection of hypotheses when true, is to relate to the procedure actually employed then the direct use of p^* is misleading; if r is sufficiently large, an apparently important effect will be found even if all null hypotheses hold. If p^* corresponds to the smallest of r independent tests then the appropriate probability of false rejection is

$$1 - (1 - p^*)^r. \tag{8.4}$$

In the special case this becomes $1 - 0.98^{20}$, which is approximately 0.33. That is, there is nothing particularly surprising in finding $p^* = 0.02$ when 20 independent tests are considered. In some contexts, such as discovery problems in particle physics and genome-wide association scans in genetics, very much larger values of r, possibly as high as 10^5, will arise.

If rp^* is small then the corrected p-value approximately equals the Bonferroni bound rp^*. Moreover this provides an upper limit to the appropriate adjusted significance level even if the tests are not independent.

It is important that adjustment to the significance level is required not so much because multiple tests are used but because a highly selected value is being considered. Thus in a factorial experiment several or sometimes many effects are studied from the same data. Provided each component of analysis addresses a separate research question, confidence intervals and tests need no allowance for multiple testing. It is only when the effect for detailed study is selected, in the light of the data, out of a number of possibilities that special adjustments may be needed.

Illustration: Considering the least significant finding In a study of patients suffering from lung disease that examined a possible association with the forced expiratory volume (FEV) of six somewhat similar biochemical substances, five showed highly significant effects but the sixth, X, had a p-value of only about 0.05. In considering the effect of X was there any allowance for the selection of this as the *least* significant effect? This is in contrast with the situation previously discussed. While no formal solution is proposed here, it seems on general grounds that if anything the p-value of X should be decreased rather than increased. A theoretical formulation of Bayesian type is possible but would require the quantitative specification of a number of essentially unknown features.

No adjustment is normally required when the details of a model are adjusted to make the specification more realistic, leaving the research question unchanged. Thus the relationship between y and x, intended for study by a linear regression, may better be examined by the regression of $\log y$ on $\log x$ if the data show that the latter relationship is more nearly linear.

Of course, when trying out several tests of the same issue, choosing one just because it gives the smallest p-value amounts to the misuse of a tool.

8.5.5 False discovery rate

In one approach to the analysis of a large number of p-values, they are arranged in decreasing order and all those smaller than some threshold are selected as evidence against the relevant null hypothesis. The threshold may be chosen to achieve a pre-assigned level of the false recovery rate, that is, the expected proportion of hypotheses that are selected as acceptable where in fact the null hypothesis is true. The chosen individual effects might then pass to a second phase of verification.

A disadvantage of this approach from some points of view is that, of the effects selected, some may show overwhelming evidence against the relevant null hypothesis, others being borderline. There is a parallel with the use of significance tests at a pre-assigned level, in which the threshold between acceptable and unacceptable null hypotheses is treated as rigidly defined.

8.5.6 Empirical Bayes formulation

Probably the most satisfactory formulation for dealing with large numbers of p-values involves an explicit model of their distribution. In such a model, with probability, say, θ the test statistic t has a null hypothesis distribution, say $f_0(t)$, whereas with probability $1 - \theta$ it has some alternative distribution, say, $f_1(t)$. If θ and the two densities $f_0(t)$, $f_1(t)$ are treated as known, the probability that a specific t comes from, say, the alternative distribution $f_1(t)$ can be found.

Note that this formulation distinguishes cases where the assignment is somewhat tentative from those in which it is relatively clear cut. In the most elaborate of these methods, θ, $f_0(t)$ and $f_1(t)$ are all treated as unknowns to be estimated whereas in the crudest version $f_0(t)$ is the theoretical null hypothesis distribution, for example the standard normal distribution, and $f_1(t)$ is a somewhat arbitrarily chosen displacement of that distribution, leaving only θ to be estimated from the data.

8.6 Estimates and standard errors

In many applications, especially when relatively standard models are fitted to fairly large amounts of data, the evidence is best summarized by an estimate of each parameter of subject-matter interest together with an associated standard error. Tests and confidence intervals may then be found immediately. The proviso is that the standard normal distribution provides an adequate approximation for the probability calculations involved. An advantage of this approach is not only its conciseness and clarity but the opportunity it gives an independent worker to make their own assessments. For example, an estimate t_1 with standard error s_{t1} can be compared with an independent estimate t_2 with standard error s_{t2}, perhaps from an independent source, by noting that $t_2 - t_1$ has standard error

$$\sqrt{(s_{t1}^2 + s_{t2}^2)}.$$

The comparison of correlated estimates needs some information about the dependence between the errors of estimation.

8.6.1 A final assessment

It is a characteristic of statistical analysis, although not always the most important one, that some assessment is provided of the uncertainty in the conclusions reached. As we have stressed, these assessments depend on assumptions about how the data were obtained and about the underlying data-generating process. The most important assumption is that the research questions of interest are captured directly in the model specification. After that, there are usually assumptions about the distributional form, often not critical, and assumptions about statistical independence, often much more critical, especially in systems with hierarchies of error structure when incorrect assumptions may lead to error estimates that are much too small.

Significance tests, confidence intervals and posterior distributions about single parameters summarize what the data on their own (or supplemented by an explicit prior distribution) tell us about the parameters of interest. The interpretation of collections of tests or confidence intervals needs particular care. For example, checking that the confidence intervals for two parameters, say ψ_1 and ψ_2, overlap is not a wise way to check the null hypothesis that $\psi_2 = \psi_1$. In fact, if the 95% level confidence limits only just overlap, there is strong evidence against this null hypothesis. The reason is that if the associated standard errors are roughly equal and the estimates independent then the difference between the means is four individual standard errors, or $4/\sqrt{2} = 2.8$ times the standard error of the difference. Again, it is hard to see from an appreciable number of confidence intervals obtained in an overview whether the results of different studies are or are not mutually consistent.

The pitfalls of selecting from a considerable number of tests only those that are significant at some nominal level, such as 0.05, are obvious. One possible general moral is that, so far as possible, fragmentary interpretation of portions of the data is to be avoided.

A final issue concerns the level of statistical detail to be given when conclusions are reported. Tables of estimates should preferably give also standard errors, but the reporting of p-values in the main text should be confined to situations where these are central to interpretation. The relentless recording of p-values and confidence intervals for every effect reported is more likely to irritate than to inform. Of course, much must depend on the conventions of the field of study.

Notes

Section 8.1. Cox (2006) gives a general introduction to the concepts underlying statistical inference.

Section 8.4. Tippett (1927) was probably the first to produce tables of random digits to aid empirical random sampling studies. The bootstrap approach was introduced by Efron (1979); for a thorough account, see Davison and Hinkley (1997). For a general account of simulation-based methods, see Ripley (1987) and for MCMC, see Robert and Casella (2004).

Section 8.5. For an early account of the analysis of large numbers of p-values, see Schweder and Spjøtvoll (1982). The basic idea of false discovery rates is due to Benjamini and Hochberg (1995); see also Storey (2002). For an elegant general account of methods for problems with a large number of similar parameters, see Efron (2010). For a simple approach to the interpretation of p-values, see Cox and Wong (2004). The notion of false discovery rates is similar to the notion of experiment-wise error rates extensively studied under the general heading of multiple comparisons (Scheffé, 1959). For Bonferroni's inequality, see Feller (1968).

9

Interpretation

Interpretation is here concerned with relating as deeply as is feasible the conclusions of a statistical analysis to the underlying subject-matter. Often this concerns attempts to establish causality, discussion of which is a main focus of the chapter. A more specialized aspect involves the role of statistical interaction in this setting.

9.1 Introduction

We may draw a broad distinction between two different roles of scientific investigation. One is to describe an aspect of the physical, biological, social or other world as accurately as possible within some given frame of reference. The other is to understand phenomena, typically by relating conclusions at one level of detail to processes at some deeper level.

In line with that distinction, we have made an important, if rather vague, distinction in earlier chapters between analysis and interpretation. In the latter, the subject-matter meaning and consequences of the data are emphasized, and it is obvious that specific subject-matter considerations must figure strongly and that in some contexts the process at work is intrinsically more speculative. Here we discuss some general issues.

Specific topics involve the following interrelated points.

- To what extent can we understand why the data are as they are rather than just describe patterns of variability?
- How generally applicable are such conclusions from a study?
- Given that statistical conclusions are intrinsically about aggregates, to what extent are the conclusions applicable in specific instances?
- What is meant by causal statements in the context in question and to what extent are such statements justified?
- How can the conclusions from the study be integrated best with the knowledge base of the field and what are the implications for further work?

159

The balance between description and understanding depends strongly on context but is not synonymous with a distinction between science and technology or between basic science and, for example, studies connected with public policy.

For example, studies of the possible health effects of exposure to mobile phone signals depend not only on epidemiological investigations of the occurrence of brain tumours but also on fundamental investigations of the possibility that such signals penetrate to the brain and of the effect that they might have on the performance of simple tasks, as assessed in an experimental psychology laboratory.

9.2 Statistical causality

9.2.1 Preliminaries

In traditional philosophical thinking a cause is necessary and sufficient for a subsequent effect. This notion is inappropriate for the kinds of situation contemplated in this book: some smokers do not get lung cancer while some non-smokers do. Yet some notion of causality underpins most scientific enquiry and the need to perceive some explanation of the world is fairly basic: Jung saw the three basic human needs to be for food, sex and a perception of causality.

At least until recently, use of the word *causal* in the statistical literature has been quite restricted, probably partly because of the strongly empirical emphasis of most statistical work and partly because of the cautious, some say overcautious, recognition of the difficulty of reaching clear unambiguous conclusions outside the research laboratory. While in one sense any use of words in a reasonably clearly defined way is acceptable, observation suggests that, particularly in reporting research in a health-related context, harm is often done by the uncritical reporting of conclusions with an unjustified explicit or implicit causal implication. The credibility of solidly based conclusions is undermined by a stream of often contradictory information on, for example, dietary matters, certainly justifying caution in that sphere.

For most statistical purposes an explanatory variable C, considered for simplicity to have just two possible values, 0 and 1, has a causal impact on the response Y of a set of study individuals if, for each individual:

- conceptually at least, C might have taken either of its allowable values and thus been different from the value actually observed; and

- there is evidence that, at least in an aggregate sense, Y values are obtained for $C = 1$ that are systematically different from those that would have been obtained on the same individuals had $C = 0$, other things being equal.

The definition of the word 'causal' thus involves the counterfactual notion that, for any individual, C might have been different from its measured value. As such, C cannot be an intrinsic variable in the sense in which we have used that term.

Illustration: Gender as a possible cause   In the great majority of contexts, gender is not to be considered as a possible cause. It usually makes no sense to consider what the outcome on a male would have been had that individual been female, other things being equal. Exceptions are studies of sex-change operations and of discriminatory employment practices. In the latter case it is reasonable to attempt to answer the question: how much would this person been paid if, with the same work skills, experience, education, etc., she had been male rather than female?

A central point in the definition of causality and its implementation concerns the requirement *other things being equal*. We discuss this in the next subsection. Note also that in this formulation there is no notion of an ultimate cause. That is, the cause C may itself be regarded as the outcome of some further process, typically occurring earlier in time or, say, at a deeper level biologically.

9.2.2 Causality and randomized experiments

It is convenient to start by discussing the interpretation of a randomized experiment. We have experimental units such that any two units may possibly receive different treatments. Each unit is then randomized to one of a number of possible treatments, in a design the details of which need not concern us here. Among the possible variables involved are the following:

- variables, B, measured before the instant of randomization, called here *baseline variables*;
- unobserved variables, U, defined before the instant of randomization;
- the treatment, T, applied to each unit;
- variables, I, measured and defined after randomization and before the outcome of interest, called *intermediate variables*; and
- response or outcome variables Y.

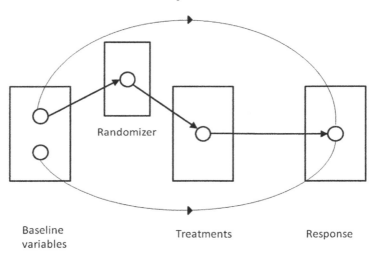

Baseline
variables   Treatments   Response

Figure 9.1 Schematic representation of a randomized
experiment. The baseline variables are defined before
randomization; the upper are observed and the lower are
unobserved. The randomizer defines treatments in such a way that
any unit is equally likely to receive each treatment, although the
precise design may depend on the observed baseline variables.
Treatment comparisons are independent of both types of baseline
variable.

See Figure 9.1.

In some contexts it may be reasonable to include in the first category
variables defined before randomization but not measured until later. The
absence of serious recall bias would be required.

The randomization used may depend on the baseline variables, notably
so if these are used to group the units into blocks for implementation of a
randomized block design.

In this formulation the requirement *other things being equal* means that
we need to contemplate the effect, for fixed B, U, of changes in the treatment. The intermediate variable I may have been affected by the treatment
and is to be ignored, that is, allowed to vary in its appropriate distribution,
in an initial analysis of Y.

In the most direct analysis any effect of B on the outcome is ignored,
except in so far as it influences the randomization scheme used; the effect of T on Y is studied directly. Any influence of U on Y is random
in such an analysis. More complicated analyses may involve a regression

analysis of Y on T and B and interpretation of the effect of T on Y for fixed B. A further development allows for an interaction between B and T. In all these analyses any effect of U is simply to inflate the random error involved.

In some very limited cases intermediate variables may be used for conditioning in such an analysis. In these cases there is sound evidence that an intermediate variable is conditionally independent of T, given B, U. Some intermediate variables may be of intrinsic interest, especially if they are predictive of the final response variable Y. Their main role, however, is likely to be in setting out pathways of dependence between the primary explanatory variables and the response.

Illustration: Adjustment for post-randomization conditions   In an investigation of a number of alternative textile spinning processes, the response variable was the end breakage rate. This was known to be affected by the relative humidity in the spinning shed, which could not be controlled. The humidity was determined by the current weather conditions and there were sound reasons for considering it to be unaffected by the process current at the time. Therefore, even though the humidity was not determined until after randomization, it was reasonable to adjust for it in analysis, that is, in effect to aim to condition the analysis on the observed relative humidity.

The intermediate variables are ignored in the first direct analysis of the effect of primary explanatory variables on the response. As mentioned above, however, the intermediate variables may themselves be response variables in further analyses. In other situations it is helpful to decompose the dependence of Y on T into a path through I, sometimes called an indirect effect, and a remainder, sometimes misleadingly called a direct effect; a better name for this remainder would be an unexplained effect. In terms of linear regression this decomposition is expressed as

$$\beta_{YT|B} = \beta_{YT|B,I} + \beta_{YI|T,B}\beta_{IT|B}. \tag{9.1}$$

Here, for example, $\beta_{YT|B,I}$ denotes the regression coefficient of Y on T in the multiple regression of Y on T, B and I, that is, I also is treated as explanatory in addition to B. The two terms on the right-hand side specify respectively the unexplained effect of T on Y and the pathway of dependence from T to I and thence to Y.

See Figure 9.2.

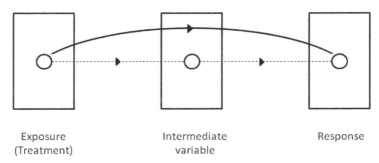

| Exposure | Intermediate | Response |
| (Treatment) | variable | |

Figure 9.2 Pathways from exposure (treatment) to response. In assessing the overall effect of exposure, the intermediate variable is ignored. The effect can be represented exactly or approximately as a combination of an indirect effect via an intermediate variable and a direct effect. The latter is the unexplained effect, that is, the one for which, as yet, no explicit intermediate explanation is available.

9.2.3 Observational parallel

In some observational studies it is a reasonable aim to parallel as far as is feasible the argument of the previous subsection. To do this the following considerations are involved.

- For any pair of variables (V, W), either V is explanatory for W, or vice versa or the two variables are to be treated as of equal standing.
- To study C as a potential cause of the behaviour of a response variable Y we examine the relationship between Y and C conditionally on all variables that are explanatory to C.
- The possibility that there is an unobserved confounder U is considered qualitatively. Such variables may be either known and unmeasured or unknown.

There are thus two sources of uncertainty in observational studies that are not present in randomized experiments. The first is that the ordering of the variables may be inappropriate, a particular hazard in cross-sectional studies.

Illustration: Interpreting cross-sectional data   In an analysis at the University of Mainz of the relationship between glucose control and patient knowledge of diabetes, the working hypothesis used was that knowledge is explanatory to success. Yet the two variables were determined at essentially the same time and it is not impossible that the incentive to learn more about the disease is influenced by the level of success achieved in controlling it.

Indeed in a sense the notion of causality is intrinsically confined to development in time. As in the previous illustration, if the data are tied to one time point then any presumption of causality relies on a working hypothesis as to whether the components are explanatory or responses. Any check on this can only be from sources external to the current data. With temporal data, models in which the values X_t, Y_t at time t are such that X_t depends on Y_t, and vice versa, should be avoided and replaced by models in which the variables at time t depend only on the variables at earlier times. There are special difficulties when, as in macroeconomics, the recorded series are heavily aggregated in time.

The second source of uncertainty is that important explanatory variables affecting both the potential cause and the outcome may not be available. In the application just mentioned it might be hoped that the relevant medical history would cover such aspects, but in many cases there may be cause for concern. Bradford Hill (1965) gave guidelines to help strengthen the case for the establishment of a causal effect in a context of interest, and in the next section we review these guidelines. Cochran (1965) reported that, when years before, he had asked R. A. Fisher about establishing causality from observational data he had received the the gnomic reply 'Make your theories elaborate', meaning that one should bring to bear as varied a set of considerations as possible; this links with some of Bradford Hill's points.

9.2.4 Qualitative guidelines

Bradford Hill's guidelines, in a form close to the original, are as follows. Evidence that an association is causal is strengthened if:

- the association is strong;
- the effect is found consistently over a number of independent studies;
- the observation of a potential cause precedes its postulated effect;
- the relationship is monotonic with respect to the level of exposure;
- the association has an underlying theoretical explanation;
- the association is based on a suitable natural experiment;
- the association is confined to the particular outcome studied.

It is to be stressed that these are guidelines or considerations, not criteria. Any or indeed all of them might be false and yet causality could still hold. Conversely, a check that they are all satisfied is no guarantee of causality.

The reason behind the first point is that the larger the effect the less likely it is to be the consequence of an unobserved confounder. Point 5 has greater cogency if the explanation is both clearly evidence-based and available beforehand. Retrospective explanations may be convincing if based on firmly established theory but otherwise need to be treated with special caution. It is well known in many fields that ingenious explanations can be constructed retrospectively for almost any finding.

In point 6, a natural experiment means a large intervention into a system, unfortunately often harmful.

Illustration: Natural interventions   Many early conclusions about HIV/ AIDS were based on studies of haemophiliacs accidently injected with contaminated blood, some becoming HIV positive. The mortality among those so infected was about 10 times that of the uninfected. Much information about the effects of radiation on health stems from H-bomb survivors from Hiroshima.

The most difficult of the guidelines to assess is the last, about the specificity of effects. In most contexts the pathways between the proposed cause and its effect are quite narrowly defined; a particular physical, biological or sociological process is involved. If this is the case then the proposed relationship should be seen as holding only in quite restricted circumstances. If the relationship in fact holds very broadly there are presumably many different processes involved all pointing in the same direction. This is not quite in the spirit of causal understanding, certainly not in the sense of the detailed understanding of an underlying process.

9.2.5 A further notion

A final definition of a causal link, probably closest to that used in the natural sciences, is that there is evidence that an underlying process has generated the data in question. This evidence may be based on well-established theory or on empirical evidence of a different kind from that immediately involved in the study under analysis, or often on a mixture of the two.

In macroscopic aspects of the physical sciences, a causal link would typically correspond to the deduction of results in classical physics or chemistry from a quantum-mechanical calculation. In the biological sciences, a causal link would correspond to an explanation in terms of knowledge at a finer scale, for example when a large-scale phenomenon is explained in terms of cellular processes. In sociology it has been suggested that explanation in terms of so-called rational action theory is causal; explanations of social regularities in terms of conclusions from psychology about individual behaviour could also be appropriately described as causal.

That this approach, applied one cautious step at a time, is often needed for deepening understanding does not imply a philosophical position of ultimate reductionism.

Illustration: Understanding the link between smoking and lung cancer
Cornfield *et al.* (2009) discussed the evidence that smoking is a cause of lung cancer. They looked at the evidence from population comparisons, from prospective and retrospective special studies, from randomized trials and from laboratory work. They examined in largely qualitative terms the alternative explanations that had been proposed and came to the conclusion that overall there was overwhelming evidence for a causal link. It is interesting that in 1959 three leading statisticians, R. A. Fisher, J. Neyman and the influential medical statistician, J. Berkson, all had reservations on the issue, even though they came from very different perspectives. It is possible that their view came in part from their assessing the evidence in a somewhat fragmentary way.

9.3 Generality and specificity

Two apparently contrasting but linked concepts concern the extension of conclusions to new situations and the applicability of conclusions from statistical analysis, inevitably concerned with aggregates, to individual situations.

In fundamental studies of new phenomena it is usual to study situations in which the aspects under study appear in their simplest and most striking form. Other considerations include the speed with which results can be obtained. In more applied contexts, the broad representativeness of the material used is one consideration but demonstration of the absence of interaction between the estimated treatment effects and key intrinsic features is probably more important.

Illustration: Stability of conclusions An important notion in the traditional approach to the design of experiments is the inclusion and planned variation of factors characterizing the experimental units, as contrasted with a very detailed study of particular situations. Thus the generality of the conclusions from an agricultural field trial would not hinge on whether the fields used were a random sample from the population of fields of interest; such a random sample would in any case not be achievable. Rather, the stability of the conclusions would depend on replication over space, in particular over soil types, and over time, representing meteorological conditions. Stability, that is the absence of any major interaction of the effects due to the spatial and temporal variables, would be the most secure base for the generality of the conclusions. The same idea, under the name of *noise factors*, is explicit in some approaches to industrial quality control.

Illustration: Determining treatment recommendations for specific patients Important medical treatments based on pharmacological products undergo a range of trials to establish their efficacy and safety. Typically they are based also on some pharmacological understanding of the mode of operation at, for example, the cellular level. Any randomized clinical trial involves patients who satisfy a formal protocol and have given informed consent; they do not form a random sample from the population of potential patients. Generality comes from repetition under a range of conditions and from biological understanding. From the perspective of a treating clinician, the issue is more that of specificity. Conclusions from randomized trials may unambiguously establish that, in aggregate, patients do better if they have all had a new treatment T than they would have done had they all received the previously standard therapy C. Clearly this is far short of showing that T is preferable to C for *every* patient, but the clinician wants to recommend the treatment appropriate for a particular patient in his or her care. The more it can be established that there is no major interaction of treatment effect with intrinsic features of patients, for example age, gender, ethnicity and previous relevant medical history, the more likely does it become that T is indeed preferable for this specific patient.

The general issue of applying conclusions from aggregate data to specific individuals is essentially that of showing that the individual does not belong to a subaggregate for which a substantially different conclusion applies. In actuality this can at most be indirectly checked for specific subaggregates.

In the particular context of a clinical trial the most direct approach is *subgroup analysis*. That is, the patients are divided into distinct subgroups to be analysed separately. This is in general a bad procedure, certainly if many subgroups are involved. On the one hand, the individual group analyses will be of low sensitivity. On the other hand, especially in the rather different situation of establishing whether there are differences in effect between T and C, the most extreme subgroup difference is likely to be quite large. As such it may be tempting to interpret it and even to forget to make an allowance for selection of the biggest effect. It is not unknown in the literature to see conclusions such as that there are no treatment differences except for males aged over 80 years, living more than 50 km south of Birmingham and life-long supporters of Aston Villa football club, who show a dramatic improvement under some treatment T. Despite the undoubted importance of this particular subgroup, virtually always such conclusions would seem to be unjustified.

A more secure approach consists of choosing a limited number of intrinsic variables, preferably in advance, and examining the interaction between the treatment effects of such variables. Such tests are of reasonable sensitivity since each uses all the data. The prior assumption is in effect that, in the general context of such trials, interactions are most likely to be small and few will be important. By contrast the isolation of anomalous subgroups leaves any anomaly uncovered unexplained.

The situation is easier when fairly quick independent replication is feasible. The usual procedure is then that any potential conclusions that have a different focus from that originally intended are subject to confirmatory study. The investigator is thus freer to follow up and even report questions not contemplated in advance.

9.4 Data-generating models

Many statistical analyses are based on models that are essentially purely descriptive accounts of patterns of variability and give relatively little guidance towards a deeper interpretation. At the other extreme some analyses involve directly a model built up essentially as a description of the data-generating process. A simple example is the discussion of *Sordaria* in Section 6.3.

An important intermediate possibility is to use data built up over time to study patterns of change and development. Such studies, often called

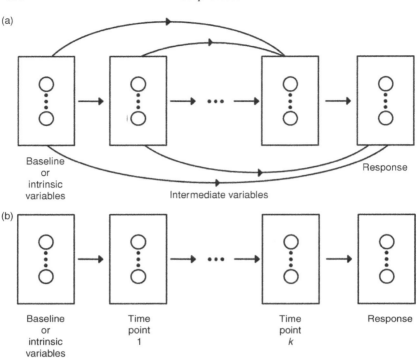

Figure 9.3 A regression chain. The variables, represented by nodes, are on an equal standing if in the same box. The variables in any box are considered conditionally on the variables in all boxes to the left. (a) A general case with arbitrary dependencies. One task of analysis is to isolate box by box the major dependencies and independencies. (b) The special case of a Markov process. The boxes correspond to time points and each box depends only on the previous box.

longitudinal, even if observational rather than experimental are such that the passage of time indicates unambiguously the direction of dependence.

This leads to the notion of a regression chain; see Figure 9.3. Here all variables in the same block are on an equal standing and commonly refer to the same time point. The variables in later boxes are responses to the variables in earlier boxes, regarded as explanatory. The idea is that by repeated regression analyses a description can be built up of a possible data-generating process. Note that no variable in any box has variables in the same box as explanatory to it; causality does not operate instantaneously! The approach is broadly comparable to the modelling of physical systems

Table 9.1 *Artificial example of interaction removed by transformation. Factor 1 and factor 2 have respectively two and three levels.*

	Factor 2		
Factor 1	Level 1	Level 2	Level 3
level 1	4	6	7
level 2	8	12	14

by differential equations based on the laws of classical physics, although there the implicit assumption is typically that the variables at any time depend only on the variables at the immediately preceding time.

9.5 Interaction

9.5.1 Preliminaries

Interaction may seem to be more a detailed issue of statistical analysis than one of interpretation, but in some contexts it has a valuable interpretive role.

First, the term interaction itself is in some ways a misnomer. There is no necessary implication of interaction in the physical sense or synergy in a biological context. Rather, interaction means a departure from additivity of the effects of two or more explanatory variables. This is expressed most explicitly by the requirement that, apart from random fluctuations, the difference in outcome between any two levels of one factor is the same at all levels of the other factor.

In Table 9.1 the factor effects combine multiplicatively rather than additively and so the interaction is removed by log transformation. In more general situations the interaction may be removable by a less familiar transformation, although if this is to an uninterpretable scale of measurement it may be best to retain the representation with interaction especially if the original scale is extensive in the sense of Section 4.3. Table 9.2 shows a complementary empirical case where the data are consistent with no interaction on an additive scale but not with a simple multiplicative effect.

The most directly interpretable form of interaction, certainly not removable by transformation, is effect reversal. This, while relatively rare in most contexts, may have strong implications. Table 9.3 shows an example where

Table 9.2   *An example where the data are*
consistent with no additive interaction but
inconsistent with no multiplicative interaction.
The data give the relative risk of lung cancer.
The risks are approximately additive but far
from multiplicative (Gustavsson et al., 2002)

	Asbestos exposure	
Smoking	No	Yes
no	1.0	10.2
yes	21.7	43.1

Table 9.3   *Qualitative interaction*
(effect reversal). The relative risk
of endometrial cancer for cyclic-
combined hormone replacement
therapy users versus non-users
classified by body mass index
(BMI) (Beral et al., 2005)

BMI	Relative risk
low	1.54
medium	1.07
high	0.57

hormone replacement therapy (HRT) appears to have beneficial or harmful consequences depending on the level of body mass index (BMI, weight in kilograms divided by the square of height in meters).

A more formal definition of interaction is as follows. Suppose that μ_{ij} is the expected response at levels i, j of the two factors under consideration. Then, the absence of two-factor interaction requires that

$$\mu_{ij} = \mu_{..} + (\mu_{i.} - \mu_{..}) + (\mu_{.j} - \mu_{..}), \tag{9.2}$$

where, for example, $\mu_{i.}$ is the average of $\mu_{i1}, \mu_{i2}, \ldots$ This implies and is implied by the requirement that the difference in response between any two levels of the first factor is the same at all levels of the second factor. In the light of this definition, the interaction components could be defined in general as the difference between the left-hand and right-hand sides of (9.2), that is by

$$\mu_{ij} - \mu_{i.} - \mu_{.j} + \mu_{..} \tag{9.3}$$

although this definition is rarely useful for direct interpretation.

We shall not discuss in detail the formal statistical procedures for testing for interaction. In principle they are based on comparing the fits of models with and without the relevant interaction.

If a representation without interaction is used as the basis for interpretation then the study of the effect of, say, the first factor would typically be based on appropriate contrasts of the mean responses at different levels, rather than directly on the comparison with an overall mean used in (9.2).

9.5.2 *Interpretation of interaction*

If an interaction is detected between the effects of two explanatory variables, subject-matter interpretation is necessary. Merely reporting the presence of interaction is rarely adequate. The approach to be used is typically different depending on whether the explanatory variables considered are

- exposures, risk factors, treatments or quasi-treatments;
- intrinsic features of the study individuals; or
- nonspecific features of the study individuals.

The primary object of study is the effect of changes in the level of the first type of explanatory variable, to be thought of in a randomized experiment as a treatment or, in an observational study, what would be a treatment were an experiment feasible. Intrinsic features characterize individuals and are not subject to hypothetical change. Nonspecific features are intrinsic features that are ill-specified or multiply specified, such as countries or centres in a multi-centre clinical trial. In the latter case, for example, there may be many ways in which centres and the patients at them may differ systematically from centre to centre and it may be neither necessary nor possible to specify these differences with any precision.

Our interest is in the interaction between an exposure and another factor. There are thus three cases to consider.

Exposure–exposure interactions where the factor levels are quantitatively specified are usually best discussed in terms of the response surface relating the expected response to the explanatory variables in question. In the relatively simple case where the surface can be represented by a second-degree polynomial, interaction corresponds to a cross-product term. The transformation of variables may be helpful. In any case interest is typically focused on the form of the response surface rather than on the interpretation of individual parameters. If one factor has quantitative levels and the other has qualitative levels then the most fruitful approach is likely to be to summarize the response to the quantitative factor, for example by slope, slope and curvature, or slope and asymptote, and to then specify how these characteristics change with the level of the qualitative factor. If both factors have unordered qualitative levels then a more *ad hoc* approach is needed. Just one of many possibilities is that most factor combinations give essentially the same mean response with only a few discrepant combinations. Caution is needed in interpreting any such patterns that are uncovered only retrospectively. If interaction is present, the study of the effects of one factor averaged over the levels of others may not be useful.

Illustration: Suitability of the average effect   Suppose that, in an experiment comparing a number of diets fed to pigs, equal numbers of males and females are used. An interaction between diets and gender means that the differences between diets are not the same for males and for females, justifying separate interpretations of inter-diet differences for the two genders. The parameters implicit in (9.2) are average effects over the two genders and these might be a useful quick summary if the interactive effect were relatively small. The average effect would be a totally appropriate summary if and only if it were necessary to make a practical recommendation for feeding the same diet to an equal mix of male and female pigs.

For an exposure–intrinsic interaction, the asymmetry of the two factors means that we may concentrate on summarizing the effect of the exposure separately at each level of the intrinsic factor. For example, the pattern of treatment effects for males may be different from the pattern for females.

Finally, an exposure–nonspecific interaction should, if possible, be given a direct interpretation via any specific information that is available but otherwise treated as a random component inducing error in the primary exposure comparisons of main interest. For example, suppose that two treatments are compared in a number of different centres or countries, that, as would often be anticipated, the treatment effect is not the same at all centres and that analysis confirms that indeed there is a treatment–centre interaction. It may be that this interaction can be described adequately by one or more features of the centres, for example their geographical location in predominantly urban or rural areas. Any such explanation is likely to be provisional. The alternative is to suppose that the interaction is generated by an effectively totally random process characterized by a component of variance. This component inflates the estimated variance of the treatment effect, the latter being defined implicitly as an average over an ensemble of repetitions of the interaction process. Note that this is not the same as regarding the centres themselves as a random sample from an explicit population of centres, typically a wholly unrealistic fiction.

9.5.3 Interaction in contingency tables

When all the variables under consideration are nominal, that is their different possible values are described qualitatively, or possibly ordinally, the data may be regarded as a contingency table of cell frequencies. For example, with three variables W, X, Y we may write

$$P(W = i, X = j, Y = k) = p_{ijk}, \tag{9.4}$$

and the corresponding observations form a three-dimensional array. Summation over each index in turn gives three two-dimensional tables summarizing the two-dimensional distributions of respectively $(X, Y), (Y, W)$ and (W, X), and further summation yields the one-dimensional marginal frequency distributions.

The analysis of such tables should follow the general approach we have discussed previously. In particular, if, as is often the case, one variable, Y, say, is best regarded as a response variable and the others as explanatory then emphasis should be placed on the form of the conditional distribution of Y given the other two variables. Such a dependence may or may not involve an interaction between X and Z on, say, the logistic scale.

If, however, three or more variables are to be treated symmetrically then there is some need to simplify the form of the joint distribution. With just two variables independence can be represented by an additive, that is, no-interaction, model on a log scale:

$$\log p_{ij} = \log p_{i.} + \log p_{.j}. \tag{9.5}$$

In a general representation we would have

$$\log p_{ij} = \log p_{i.} + \log p_{.j} + \xi_{ij}, \tag{9.6}$$

say. For a closer parallel with representations for continuous variables, we rewrite (9.6) in the form

$$\log p_{ij} = \mu + \alpha_{i.} + \alpha_{.j} + \gamma_{ij}, \tag{9.7}$$

where, for example,

$$\alpha_{i.} = \text{ave}_j \log p_{ij} - \text{ave}_{i,j} \log p_{ij}. \tag{9.8}$$

No interaction, that is, the vanishing of all the γ_{ij}, is another way of expressing the independence of the two variables, conventionally denoted by $X \perp\!\!\!\perp Y$.

For three or more variables, care is needed in interpreting the conditional as contrasted with the marginal independence of variables, for example the marginal independence $X \perp\!\!\!\perp Y$ as contrasted with the conditional independence $X \perp\!\!\!\perp Y \mid Z$. In fact unless $(X, Y) \perp\!\!\!\perp Z$, that is, the combined variable is independent of Z, the two notions of independence will not be the same. In this context independence is expressed in terms of log probabilities, whereas marginalization involves the addition of probabilities.

Illustration: Strength of association varies between countries   In some studies of social mobility (Erikson and Goldthorpe, 1992), the interest lies in the relationship between current social class, Y, and parental social class, X, using data from a number of different countries Z. While Z is an intrinsic or even a nonspecific variable and Y and X may be reasonably be regarded as response and explanatory variables respectively, interest has focused on the empirical finding that while the interaction, that is, the lack of independence, between X and Y is different for different countries, the broad pattern is essentially constant. This can be formulated by a model in which each country has a model of the form (9.6) with separate parameters for each level k of Z, constrained only by

$$\xi_{ij}^k = \rho_k \xi_{ij}; \tag{9.9}$$

here ρ_k characterizes the strength of association in country k.

9.5.4 Three-factor and higher-order interactions

Formally, interactions of any order can be defined recursively. Thus, the absence of a three-factor interaction means that the pattern of two-factor interactions is the same at all levels of the third factor. On an additive scale this definition is symmetrical between the three factors, despite the apparent asymmetry of the definition. An explicit definition is that if μ_{ijk} denotes the expected response at levels i, j, k of the three factors then the absence of three-factor interaction requires that

$$
\begin{aligned}
\mu_{ijk} = \mu_{\ldots} &+ (\mu_{i..} - \mu_{\ldots}) + (\mu_{.j.} - \mu_{\ldots}) + (\mu_{..k} - \mu_{\ldots}) \\
&+ (\mu_{ij.} - \mu_{i..} - \mu_{.j.} + \mu_{\ldots}) + (\mu_{i.k} - \mu_{i..} - \mu_{..k} + \mu_{\ldots}) \\
&+ (\mu_{.jk} - \mu_{.j.} - \mu_{..k} + \mu_{\ldots}).
\end{aligned}
\tag{9.10}
$$

In practice, except in some contingency tables and balanced factorial experiments, three- and higher-factor interactions seem only rarely to be used directly. In balanced factorial experiments with continuous approximately normally distributed responses, in principle interactions of all orders are easily calculated and easily set out in the form of an analysis of variance table. The distinction between treatment factors and intrinsic factors defining the experimental units remains important for interpretation.

One general point that can emerge from such an analysis of variance is that if there are many interactions involving a particular factor, especially an intrinsic feature, then it is likely that there are quite different patterns of response at the different levels and that the data and analysis are best split into parts. Thus it may be better to analyse data on males and on females separately if many appreciable interactions involving gender appear in the analysis.

Another possibility is that nearly all the factor combinations have essentially the same response, only one or a small number showing a clearly different outcome from the majority. Then interactions of all orders are likely to be appreciable, revealing that the chosen representation in terms of overall factor effects and interactions is inappropriate.

9.6 Temporal data

9.6.1 Types of data

The study of progression in time is often a key to understanding. Special statistical methods for such data fall into at least two broad types.

The first is based on the availability of quite long series of data behaving in a relatively stable way. Typical examples are meteorological data, for example on hourly, daily, monthly, annual, etc. rainfall, and some kinds of neurological data recording electrical signals between cells.

Somewhat similar methods apply to the analysis of spatial and spatial–temporal data but there is a crucial distinction between time and space: the arrow of time has a uniquely defined direction!

The second kind of data are relatively short series collected on a large number of essentially independent individuals. Thus a panel of individuals may be interviewed every year for a few years to enquire about their voting intentions, attitudes to key political and social issues and so on. In another type of investigation, following injection of a monitoring substance the blood concentration may be recorded at regular intervals for, say, a 24 hr period. The object is to assess the resulting concentration-time response curves, possibly in order to compare different groups of individuals.

Both types of data lead to a wide variety of statistical methods, which we shall not attempt to summarize. Instead we give a few broad principles that may help an initial approach to such data. An important general precept applying to both types of data, and indeed more generally, is that qualitative inspection of the data, or of a sample of it, is crucial before more formal methods of analysis are applied.

9.6.2 *Time series analysis*

A central notion in time series analysis is that of time scales of variation. These may take various forms, for example local statistical dependencies, periodic fluctuations of known wavelength and slow trends or drifts. These types of variation should be decoupled as far as possible before detailed analysis if confusion is to be avoided.

The sampling interval, that is, the spacing between successive observations, plays an important role. Moreover, if, say, daily temperature data are being analysed then it needs to be clear whether the individual values are point values at some specified time of day or are in some sense aggregated or averaged values over a period.

Short-term haphazard variation is usually best studied by a series of plots of the value at time t against first the preceding value y_{t-1} and then $y_{t-2}, y_{t-3}, \ldots$ Each value of t gives one point on each plot. If the relationships are reasonably linear, the information may then be summarized in serial correlation coefficients, $r_1, r_2, \ldots$, that is, the standard correlation coefficients from the above plots. Analyses of this broad type, including the study of lagged relationships between different variables, are said to be in the *time domain*. The correlation pattern may be used essentially descriptively or as a basis for fitting formal models typically representing the relationship between the value at time t and previous values.

The use of correlations implies a restriction to essentially linear relations. Nonlinear time series models are of special interest as examples of the sometimes exotic behaviour of nonlinear systems.

Methods based on correlations or covariances may be contrasted with the *frequency domain* approach, which is popular for good reasons, especially in the physical sciences. In this approach, for n equally spaced observations at unit time spacing, a series of sinusoidal fluctuations is considered that span the range from very rapid to very slow change. These have the form

$$\cos(\omega_p t), \quad \sin(\omega_p t), \tag{9.11}$$

where the values of $\omega_p = 2\pi p/n$ span the interval $(0, 2\pi)$. An analysis of variance is now possible showing how much of the variance in the data is assigned to the various ranges of ω. While conceptually this is quite different from time domain analysis, mathematically and numerically it is essentially equivalent, both approaches being based on a large number of quadratic functions of the data. When there is no strict periodicity in the data, a smoothed version of the analysis of variance gives the power spectral density function of the data, in effect showing how the power, synonymous here with the variance, is distributed across the range from small ω, that is, long-term variation, to large ω, that is, short-term variation. For totally random data the power spectrum is flat and the term *white noise* is therefore used for such data.

A clear periodic fluctuation of known wavelength, such as a dependence on the time of day or season of the year, will be shown by strong contributions at the relevant values of the frequency ω or associated wavelength $2\pi/\omega$ but is better handled by a more elementary method. In the latter we reorganize the data as a two-way table in which, for example, the columns correspond to the time of day at which data are available and the rows to days. Marginal means and residuals may now be found and interpreted.

More elaborate methods suitable for studying some kinds of local behaviour use a decomposition not into sinusoidal components but into wavelets.

9.6.3 Longitudinal data

There are broadly three main approaches to the analysis of longitudinal data, any one or combination of which may be needed. We consider the simplest case where data are available on a number of independent individuals. The three approaches are as follows.

- Summarize the data on each individual by a few statistics chosen on general grounds and treat these as initial variables in an analysis concentrating on comparisons *between* individuals.

- Analyse the data in two phases, first fitting a suitable model to the data for each individual and then in a second phase analysing inter-individual variation.
- Fit a single *multi-level* model in which all variation is represented consistently in a single form.

The choice clearly depends on the focus of interest. We have already discussed the first approach in connection with the illustration 'Summarizing response curves' (p. 115) and so now we discuss briefly just the second and third possibilities.

Probably the simplest possibility is that for each individual there are observations of a response y at a series of time points and that a broadly linear relationship holds for each individual. If the time points are fairly widely spaced then a linear regression model for individual i,

$$y_{it} = \alpha_i + \beta_i(t - t_0) + \epsilon_{it}, \tag{9.12}$$

with independent errors ϵ_{it} might be taken; otherwise time series models for the correlations between errors at different time points might be needed.

In the simple version, for the first stage of analysis we would have an estimated linear regression coefficient b_i and an intercept a_i, the latter calculated at the reference time point t_0 common to all individuals. In the second phase of analysis the estimates (a_i, b_i) are studied. How variable are they? Are they related to other observed features of the individuals? In a study in which the data for all individuals are taken at the same set of time points the (a_i, b_i) will be of equal precision, but in many applications the configuration of time points will not be the same and any need to allow for differing precisions may complicate this essentially simple procedure.

In the third approach we use a multi-level model in which

$$a_i = \mu_a + \eta_i, \qquad b_i = \mu_b + \zeta_i, \tag{9.13}$$

where η_i and ζ_i are random terms of mean zero having an unknown covariance matrix that is independent of ϵ_i. The representation (9.13) can be augmented, for example by terms representing systematic dependencies on explanatory variables defined at the individual level. The random variation is now described by three variances and a covariance; software is available for estimating the full model (9.12), (9.13).

9.7 Publication bias

Publication bias is most commonly thought of in simple terms: statistically significant, or positive, study results are more likely to be published.

The effect, recognized for many years (Sterling, 1959; Dawid and Dickey, 1977), was labelled the "file drawer effect" in 1979 (Rosenthal, 1979) and, although files are now more likely to be computer- rather than drawer-based, the phenomenon remains. There are other related biases that make significant results more readily available than their non-significant counterparts. The Cochrane Collaboration (no date) listed four additional biases: statistically significant results are more likely to be published rapidly (time-lag bias); they are more likely to be published in English (language bias); they are more likely to be published more than once (multiple publication bias); and they are more likely to be cited (citation bias). Furthermore, significant results are more likely to be published in journals with high impact factors. These effects have been well documented in the clinical literature but they have also been examined in areas as diverse as studies of political behaviour (Gerber *et al.*, 2010), of the potentially violent effects of video games (Ferguson, 2007) and of climate change (Michaels, 2008).

Efforts to reduce such biases include trial registries, such as the US clinicaltrials.gov, which aim to document all clinical trials undertaken regardless of the significance or direction of their findings. Funnel plots are often used to detect publication bias in studies bringing together results from previously published studies. The aim of these is to recognize asymmetry in the published results such as might arise from publication bias, as well as from time-lag bias and, potentially, from English-language bias depending on the form of the literature search. Regression methods are sometimes available to adjust for the effects of publication bias.

9.8 Presentation of results which inform public policy

Particular challenges arise when results of analyses which directly influence debate and decisions on public policy are to be presented. There may be considerable pressure to provide results which can be seen as justifying a particular decision or for a level of precision which indicates the unambiguous superiority of one policy over all others under consideration. While the broad principles for explaining policy-relevant scientific results are no different from those for any other scientific results (see Section 5.4 for points regarding the graphical presentation of data and results), the risks of misinterpretation, deliberate or otherwise, may well be greater if difficult decisions hinge in part on the results.

Thus, it is critical that in presenting results scientists are careful to ensure that:

- where necessary, important conclusions are explained directly to key decision makers, not through intermediaries;
- possibly unwelcome aspects of the conclusions are not suppressed or hidden;
- the underlying assumptions are clearly set out;
- key conclusions are given with some indication of uncertainty;
- the extent to which conclusions change if the analytical approach is varied is explained;
- results are placed in the context of information from other relevant studies;
- the representativeness of the data under study for the conditions under which the conclusions are to be applied is discussed; and
- the data underlying the analysis, as well as the basis of the analysis, are made publicly available. for example by providing the raw data as supplementary material published with a paper in the subject-matter scientific literature.

With regard to uncertainty, there may well be contexts where two different analyses yield different estimates of precision. The results may be naively interpreted as implying that the analysis yielding the greater certainty (as reflected by a narrower confidence interval or lower p-value) is better, and it is a particular challenge to explain clearly to non-specialists, in particular non-scientists, which is the more appropriate analysis and why.

Illustration: Disease control policy based on multiple independent epidemiological analyses   In 2001 there was a large outbreak of foot-and-mouth disease (FMD), a disease of cloven-hooved mammals, in the UK. The UK had been clear of FMD for many years and thus the disease spread rapidly amongst the fully susceptible populations of cattle, sheep and pigs. The then Chief Scientific Advisor, Sir David King, brought together independent groups of epidemiologists to analyse data on the spread of the disease and on the implementation of disease control through the slaughtering of animals in herds or flocks in which infection had been detected. Independent analyses conducted by groups at Imperial College London, the University of Cambridge and the University of Edinburgh all indicated that the original test-confirm-and-slaughter policy was insufficient to control the epidemic, and on the basis of these consistent results the Chief Scientific Advisor recommended that a more ambitious control policy be implemented. This policy aimed to slaughter animals from infected herds or flocks within 24 hours of detection and to slaughter animals from herds or flocks contiguous to those infected within 48 hours of the detection. The epidemic was brought under control following the implementation of this policy, although whether other policies might have been equally effective is still debated.

Notes

Section 9.5. For a general discussion of interaction, see Cox (1984) and Berrington de Gonzáles and Cox (2007). The analysis of variance is described at length in some older books on statistical methods, for example Snedecor and Cochran (1956).

Section 9.6. For methods for time series analysis see Brockwell *et al.* (2009), for longitudinal data see Diggle *et al.* (2002) and for multi-level modelling see Snijders and Bosker (2011). There are extensive literatures on all these topics.

10

Epilogue

In this chapter we deal with various issues which evaded discussion in earlier chapters and conclude with a broad summary of the strategical and tactical aspects of statistical analysis.

10.1 Historical development

The development of new statistical methods stems directly or indirectly from specific applications or groups of related applications. It is part of the role of statistical theory in the broad sense not only to develop new concepts and procedures but also to consolidate new developments, often developed in an *ad hoc* way, into a coherent framework. While many fields of study have been involved, at certain times specific subjects have been particularly influential. Thus in the period 1920–1939 there was a strong emphasis on agricultural research; from 1945 problems from the process industries drove many statistical developments; from 1970 onwards much statistical motivation has arisen out of medical statistics and epidemiology. Recently genetics, particle physics and astrophysics have raised major issues. Throughout this time social statistics, which gave the subject its name, have remained a source of challenge.

Clearly a major force in current developments is the spectacular growth in computing power in analysis, in data storage and capture and in highly sophisticated measurement procedures. The availability of data too extensive to analyse is, however, by no means a new phenomenon. Fifty or more years ago the extensive data recorded on paper tapes could scarcely be analysed numerically, although the now-defunct skill of analogue computation could occasionally be deployed.

Currently, many areas of the sciences and associated technologies involve statistical issues and often statistical challenges; some have been mentioned in previous chapters. An interesting group of ideas, largely from a computer science context, comes under the broad names *data mining* and *machine learning*. The characteristics of some, although not all, such applications are the following.

- Large or very large amounts of data are involved.
- The primary object is typically empirical prediction, usually assessed by setting aside a portion of the data for independent checks of the success of empirical formulae, often judged by mean square prediction error.
- Often there is no explicit research question, at least initially.
- There is little explicit interest in interpretation or in data-generating processes, as contrasted with prediction.
- Any statistical assessment of errors of estimation usually involves strong independence assumptions and is likely to seriously underestimate real errors.
- In data mining there is relatively little explicit use of probability models.
- Methods of analysis are specified by algorithms, often involving numerical optimization of plausible criteria, rather than by their statistical properties.
- There is little explicit discussion of data quality.

A broader issue, especially in machine learning, is the desire for wholly, or largely, automatic procedures. An instance is a preference for neural nets over the careful semi-exploratory use of logistic regression. This preference for automatic procedures is in partial contrast with the broad approach of the present book. Here, once a model has been specified, the analysis is largely automatic and in a sense objective, as indeed is desirable. Sometimes the whole procedure is then relatively straightforward. We have chosen, however, to emphasize the need for care in formulation and interpretation.

Breiman (2001) argued eloquently that algorithmic-based methods tend to be more flexible and are to be preferred to the mechanical use of standard techniques such as least squares or logistic regression.

10.2 Some strategic issues

We discuss here a few general points that may arise at any stage of an investigation but are probably of most relevance to the analysis of data.

So far as feasible, analyses should be transparent in the sense that one can see in outline the pathways between the data and the conclusions. Black-box methods, in which data are fed into a complex computer program which emits answers, may be unavoidable but should be subject to informal checks.

Sometimes data arise in a number of sections, each with a viable analysis. It may then be best, partly for the reason sketched in the previous paragraph, to analyse the sections separately and then, as appropriate, synthesize the separate answers. The alternative of a single analysis with checks for interactions between sections and key features may in a sense be more efficient but is likely to be less secure and certainly less transparent. Thus in a large social or medical study it may be best first to analyse the data for men and for women separately. The alternative is a single model with interactions with gender as judged appropriate. If, however, the component parts of such an analysis contain little information then a single analysis will be preferable.

It has to be recognized that approximations are involved in all thinking, quantitative or qualitative, about complicated situations. The approximations are explicit in formal statistical analysis through the specification of a model. Judgement has to be exercised in how simple a model may be used, but it is often a good strategy to start with a rather simple model and then to ask the question 'What assumptions have been made that might seriously affect the conclusions?' It may or may not be clear that no change in the model assumptions is likely seriously to change the conclusions.

To deal with this last issue, it is desirable that models are set out explicitly either in mathematical notation or in words. In particular, independence assumptions about random terms should be specified. The tempting practice of specifying a model solely by the computer package used to fit it should be shunned. It hides the assumptions implicit in the analysis.

It is essential with large investigations that a plan of analysis is formulated at the design stage, in principle flexible enough to deal with the unexpected. While there may be pressures to follow it with some rigidity, it is likely that some minor, or maybe not so minor, changes are essential in the light of experience gained in collecting the data and during analysis.

10.3 Some tactical considerations

We now turn to some more detailed but still broadly applicable issues of statistical analysis.

The key features of any statistical model used as a base for analysis and interpretation are, in decreasing order of general importance, that the form of systematic variation is well specified, that independence assumptions about random terms are appropriate and that any assumptions about the parametric forms of probability distributions are reasonable. A particular issue concerning large amounts of data is that uncritical assumptions about statistical independence may lead to a gross underestimate of the real error in conclusions.

It is important that the main stages in the output of statistical analyses are properly annotated, so that the analyses are readily understandable at some unspecified time in the future. Even though purely numerical analytic flaws in analysis due to rounding-off errors are unlikely, mistakes can arise, and it is important to have ways of checking that answers are 'reasonable'; even some old-fashioned facility at mental arithmetic is not to be despised.

The essential principles are as follows.

- *Study design* should aim to give clear answers to focused questions, should be of appropriate size and have at least an outline scheme of analysis.
- The intended role in interpretation of *different kinds of measurement* should be clear.
- The measurements should be of *appropriate quality* and this quality should be monitored directly or indirectly.
- Both the preliminary and more formal analysis should be as *simple and transparent* as possible.
- *Conclusions* of a statistical analysis should be clearly stated in a form linked as directly as possible to the subject-matter conclusions. The sensitivity of the conclusions to any simplifications made in the analysis should be assessed.

Flexible formulation of the research questions at issue is important so that later statistical analysis addresses and quite possibly modifies and clarifies those questions. The requirements may range from obtaining secure information about the state of the world as it is to the more speculative task of understanding underlying processes. The natural time scale of an investigation, roughly the average time taken from conception of the idea to the summarization of conclusions, has a great bearing on how such investigations should proceed.

10.4 Conclusion

In the light of the title of the book, the reader may reasonably ask: What then really are the principles of applied statistics? Or, more sceptically, and equally reasonably: in the light of the great variety of current and potential applications of statistical ideas, can there possibly be any universal principles?

It is clear that any such principles can be no more than broad guidelines on how to approach issues with a statistical content; they cannot be prescriptive about how such issues are to be resolved in detail.

The overriding general principle, difficult to achieve, is that there should be a seamless flow between statistical and subject-matter considerations.

This flow should extend to all phases of what we have called the ideal sequence: formulation, design, measurement, the phases of analysis, the presentation of conclusions and finally their interpretation, bearing in mind that these phases perhaps only rarely arise in such a sharply defined sequence. To put this in slightly more personal terms, in principle seamlessness requires an individual statistician to have views on subject-matter interpretation and subject-matter specialists to be interested in issues of statistical analysis.

No doubt often a rather idealized state of affairs. But surely something to aim for!

References

Ahern, M. J., Hall, N. D., Case, K., and Maddison, P. J. (1984). D-penicillamine withdrawal in rheumatoid arthritis. *Ann. Rheum. Dis.* **43**, 213–217. (Cited on p. 85.)

Amemiya, T. (1985). *Advanced Econometrics*. Harvard University Press. (Cited on p. 52.)

American Statistical Association Committee on Professional Ethics (1999). Ethical guidelines for statistical practice. (Cited on p. 75.)

Baddeley, A., and Jensen, E. B. V. (2005). *Stereology for Statisticians*. Chapman and Hall. (Cited on pp. 51 and 91.)

Bailey, R. A. (2008). *Design of Comparative Experiments*. Cambridge University Press. (Cited on p. 51.)

Ben-Akiva, M., and Leman, S. R. (1985). *Discrete Choice Analysis: Theory and Application to Travel Demand*. MIT Press. (Cited on p. 49.)

Benjamini, Y., and Hochberg, Y. (1995). Controlling the false discovery rate: a practical and powerful approach to multiple testing. *J. R. Statist. Soc. B* **57**, 289–300. (Cited on p. 158.)

Beral, V., Bull, D., Reeves, G., and Million Women Study Collaborators (2005). Endometrial cancer and hormone-replacement therapy in the Million Women Study. *Lancet* **365**, 1543–1551. (Cited on p. 172.)

Berrington de Gonzáles, A., and Cox, D. R. (2007). Interpretation of interaction: a review. *Ann. Appl. Statist.* **1**, 371–385. (Cited on p. 183.)

Beveridge, W. I. B. (1950). *The Art of Scientific Investigation*. First edition. Heinemann, (Third edition published in 1957 by Heinemann and reprinted in 1979.) (Cited on p. 13.)

Bird, S. M., Cox, D., Farewell, V. T., Goldstein, H., Holt, T., and Smith, P. C. (2005). Performance indicators: good, bad, and ugly. *J. R. Statist. Soc. A* **168**, 1–27. (Cited on p. 55.)

Bollen, K. A. (1989). *Structural Equations with Latent Variables*. Wiley-Interscience. (Cited on p. 74.)

Booth, K. H. V., and Cox, D. R. (1962). Some systematic supersaturated designs. *Technometrics* **4**, 489–495. (Cited on p. 138.)

Bourne, J., Donnelly, C., Cox, D., *et al.* (2007). Bovine TB: the scientific evidence. A science base for a sustainable policy to control TB in cattle. Final report of the independent scientific group on cattle TB. Defra. (Cited on p. 17.)

Box, G. E. P. (1976). Science and statistics. *J. Am. Statist. Assoc.* **71**, 791–799. (Cited on p. 13.)

Box, G. E. P., Hunter, W. G., and Hunter, J. S. (1978). *Statistics for Experimenters: Design, Innovation, and Discovery.* John Wiley & Sons. (Second edition published in 2005 by Wiley-Interscience.) (Cited on p. 51.)

Breiman, L. (2001). Statistical modeling: the two cultures. *Statist. Sci.* **16**, 199–231. (Cited on p. 185.)

Breslow, N. E., and Day, N. E. (1980). *Statistical Methods in Cancer Research. Vol. 1: The Analysis of Case-Control Studies.* International Agency for Research on Cancer. (Cited on p. 52.)

Breslow, N. E., and Day, N. E. (1987). *Statistical Methods in Cancer Research. Vol. II: The Design and Analysis of Cohort Studies.* International Agency for Research on Cancer. (Cited on p. 52.)

Brockwell, P., Fienberg, S. E., and Davis, R. A. (2009). *Time Series: Theory and Methods.* Springer. (Cited on p. 183.)

Brody, H., Rip, M. R., Vinten-Johansen, P., Paneth, N., and Rachman, S. (2000). Mapmaking and myth-making in Broad Street: the London cholera epidemic, (1854). *Lancet* **356**(07), 64–68. (Cited on p. 82.)

Bunting, C., Chan, T. W., Goldthorpe, J., Keaney, E., and Oskala, A. (2008). *From Indifference to Enthusiasm: Patterns of Arts Attendance in England.* Arts Council England. (Cited on p. 68.)

Büttner, T., and Rässler, S. (2008). Multiple imputation of right-censored wages in the German IAB Employment Sample considering heteroscedasticity. IAB discussion paper 2000 844. (Cited on p. 62.)

Carpenter, L. M., Maconochie, N. E. S., Roman, E., and Cox, D. R. (1997). Examining associations between occupation and health using routinely collected data. *J. R. Statist. Soc. A* **160**, 507–521. (Cited on p. 4.)

Carpenter, L. M., Linsell, L., Brooks, C., *et al.* (2009). Cancer morbidity in British military veterans included in chemical warfare agent experiments at Porton Down: cohort study. *Br. Med. J.* **338**, 754–757. (Cited on p. 48.)

Carroll, R. J., Ruppert, D., Stefanski, L. A., and Crainiceanu, C. M. (2006). *Measurement Error in Nonlinear Models: A Modern Perspective.* Second edition. Chapman & Hall/CRC. (Cited on p. 74.)

Chalmers, A. F. (1999). *What is This Thing Called Science?* Third edition. Open University Press. (Cited on p. 13.)

Champion, D. J., and Sear, A. M. (1969). Questionnaire response rate: a methodological analysis. *Soc. Forces* **47**, 335–339. (Cited on p. 54.)

Chatfield, C. (1998). *Problem-solving.* Second edition. Chapman and Hall. (Cited on p. 13.)

Choy, S. L., O'Leary, R., and Mengersen, K. (2009). Elicitation by design in ecology: using expert opinion to inform priors for Bayesian statistical models. *Ecology* **90**, 265–277. (Cited on p. 143.)

Cochran, W. G. (1965). The planning of observational studies of human populations (with discussion). *J. R. Statist. Soc. A* **128**, 234–266. (Cited on p. 165.)

Cochrane Collaboration. What is publication bias? http://www.cochrane-net.org/openlearning/HTML/mod15-2.htm.

Coleman, J. S., and James, J. (1961). The equilibrium size distribution of freely-forming groups. *Sociometry* **24**, 36–45. (Cited on p. 102.)

Cornfield, J., Haenszel, W., Hammond, E. C., Lilienfeld, A. M., Shimkin, M. B., and Wynder, E. L. (2009). Smoking and lung cancer: recent evidence and a discussion of some questions (reprinted from (1959). *J. Nat. Cancer Inst.* **22**, 173–203). *Int. J. Epidemiol.* **38**, 1175–1191. (Cited on p. 167.)

Cox, D. R. (1952). Some recent work on systematic experimental designs. *J. R. Statist. Soc. B* **14**, 211–219. (Cited on p. 51.)

Cox, D. R. (1958). *Planning of Experiments*. John Wiley & Sons. (Cited on p. 51.)

Cox, D. R. (1969). Some sampling problems in technology in *New Developments in Survey Sampling*, pages 506–527. John Wiley and Sons. (Cited on p. 51.)

Cox, D. R. (1984). Interaction. *International Statist. Rev.* **52**, 1–31. (Cited on p. 183.)

Cox, D. R. (1990). Role of models in statistical analysis. *Statist. Sci.* **5**, 169–174. (Cited on p. 117.)

Cox, D. R. (2006). *Principles of Statistical Inference*. Cambridge University Press. (Cited on p. 158.)

Cox, D. R., and Brandwood, L. (1959). On a discriminatory problem connected with the works of Plato. *J. R. Statist. Soc. B* **21**, 195–200. (Cited on p. 105.)

Cox, D. R., and Oakes, D. (1984). *Analysis of Survival Data*. Chapman and Hall. (Cited on p. 139.)

Cox, D. R., and Reid, N. (2000). *Theory of the Design of Experiments*. Chapman and Hall. (Cited on pp. 51 and 139.)

Cox, D. R., and Snell, E. J. (1979). On sampling and the estimation of rare errors. *Biometrika* **66**, 125–132. (Cited on p. 51.)

Cox, D. R., and Wermuth, N. (1996). *Multivariate Dependencies*. Chapman and Hall. (Cited on pp. 47, 58, and 103.)

Cox, D. R., and Wong, M. Y. (2004). A simple procedure for the selection of significant effects. *J. R. Statist. Soc. B* **66**, 395–400. (Cited on p. 158.)

Cox, D. R., Fitzpatrick, R., Fletcher, A., Gore, S. M., Spiegelhalter, D. J., and Jones, D. R. (1992). Quality-of-life assessment: can we keep it simple? (with discussion). *J. R. Statist. Soc. A* **155**, 353–393. (Cited on p. 74.)

Darby, S., Hill, D., Auvinen, A., *et al.* (2005). Radon in homes and risk of lung cancer: collaborative analysis of individual data from 13 European case-control studies. *Br. Med. J.* **330**, 223–227. (Cited on p. 50.)

Davison, A. C., and Hinkley, D. V. (1997). *Bootstrap Methods and Their Application*. Cambridge University Press. (Cited on p. 158.)

Dawid, D. P., and Dickey, J. M. (1977). Likelihood and Bayesian inference from selectively reported data. *J. Am. Statist. Assoc.* **72**, 845–850. (Cited on p. 181.)

de Almeida, M. V., de Paula, H. M. G., and Tavora, R. S. (2006). Observer effects on the behavior of non-habituated wild living marmosets (*Callithrix jacchus*). *Rev. Etologia* **8**(2), 81–87. (Cited on p. 54.)

De Silva, M., and Hazleman, B. L. (1981). Long-term azathioprine in rheumatoid arthritis: a double-blind study. *Ann. Rheum. Dis.* **40**, 560–563. (Cited on p. 85.)

Diggle, P., Liang, K.-Y., and Zeger, S. L. (2002). *Analysis of Longitudinal Data*. Second edition. Oxford University Press. (Cited on p. 183.)

Donnelly, C. A., Woodroffe, R., Cox, D. R., *et al.* (2003). Impact of localized badger culling on tuberculosis incidence in British cattle. *Nature* **426**, 834–837. (Cited on p. 17.)

Donnelly, C. A., Woodroffe, R., Cox, D. R., *et al.* (2006). Positive and negative effects of widespread badger culling on tuberculosis in cattle. *Nature* **439**, 843–846. (Cited on p. 17.)

Edge, M. E., and Sampaio, P. R. F. (2009). A survey of signature based methods for financial fraud detection. *Comput. Secur.* **28**, 381–394. (Cited on p. 79.)

Efron, B. (1979). Bootstrap methods: another look at the jackknife. *Ann. Statist.* **7**, 1–26. (Cited on pp. 149 and 158.)

Efron, B. (2010). *Large-Scale Inference.* IMS Monograph. Cambridge University Press. (Cited on p. 158.)

Erikson, R., and Goldthorpe, J. H. (1992). *The Constant Flux.* Oxford University Press. (Cited on p. 176.)

Feller, W. (1968). *An Introduction to Probability Theory and its Applications. Vol. 1.* Third edition. Wiley. (Cited on p. 158.)

Ferguson, C. J. (2007). The good, the bad and the ugly: a meta-analytic review of positive and negative effects of violent video games. *Psychiat. Quart.* **78**, 309–316. (Cited on p. 181.)

Firth, D., and de Menezes, R. X. (2004). Quasi-variances. *Biometrika* **91**, 65–80. (Cited on p. 139.)

Fisher, R. A. (1926). The arrangement of field experiments. *J. Min. Agric. G. Br.* **33**, 503–513. (Cited on pp. 28 and 51.)

Fisher, R. A. (1935). *Design of Experiments.* Oliver and Boyd. (Cited on pp. 28 and 51.)

Fleming, T. R., and DeMets, D. L. (1996). Surrogate end points in clinical trials: are we being misled? *Ann. Intern. Med.* **125**, 605–613. (Cited on p. 60.)

Forrester, M. L., Pettitt, A. N., and Gibson, G. J. (2007). Bayesian inference of hospital-acquired infectious diseases and control measures given imperfect surveillance data. *Biostatistics* **8**, 383–401. (Cited on pp. 100 and 101.)

Galesic, M., and Bosnjak, M. (2009). Effects of questionnaire length on participation and indicators of response quality in a web survey. *Public Opin. Quart.*, **73**, 349–360. (Cited on p. 54.)

Gerber, A. S., Malhotra, N., Dowling, C. M., and Doherty, D. (2010). Publication bias in two political behavior literatures. *Am. Polit. Research* **38**, 591–613. (Cited on p. 181.)

Gile, K., and Handcock, M. S. (2010). Respondent-driven sampling: an assessment of current methodology. *Sociol. Methodol.* **40**, 285–327. (Cited on p. 51.)

Gøtzsche, P. C., Hansen, M., Stoltenberg, M., *et al.* (1996). Randomized, placebo controlled trial of withdrawal of slow-acting antirheumatic drugs and of observer bias in rheumatoid arthritis. *Scand. J. Rheumatol.* **25**, 194–199. (Cited on p. 85.)

Grady, D., Rubin, S. H., Petitti, D. B., *et al.* (1992). Hormone therapy to prevent disease and prolong life in postmenopausal women. *Ann. Intern. Med.* **117**, 1016–1037. (Cited on p. 7.)

Greenland, S., and Robins, J. M. (1988). Conceptual problems in the definition and interpretation of attributable fractions. *Am. J. Epidemiol.* **128**, 1185–1197. (Cited on p. 120.)

Gustavsson, P., Nyberg, F., Pershagen, G., Schéele, P., Jakobsson, R., and Plato, N. (2002). Low-dose exposure to asbestos and lung cancer: dose-response relations and interaction with smoking in a population-based case-referent study in Stockholm, Sweden. *Am. J. Epidemiol.* **155**, 1016–1022. (Cited on p. 172.)

Guy, W. A. (1879). On tabular analysis. *J. Statist. Soc. Lond.* **42**, 644–662. (Cited on p. 86.)

Hand, D. J. (2004). *Measurement: Theory and Practice.* Arnold. (Cited on p. 74.)

Hand, D. J., Mannila, H., and Smyth, P. (2001). *Principles of Data Mining.* MIT Press. (Cited on p. 13.)

Herman, C. P., Polivy, J., and Silver, R. (1979). Effects of an observer on eating behavior: The induction of "sensible" eating. *J. Pers.* **47**(1), 85–99. (Cited on p. 54.)

Hill, A. B. (1965). The environment and disease: association or causation? *Proc. R. Soc. Med.* **58**, 295–300. (Cited on p. 165.)

Hogan, J. W., and Blazar, A. S. (2000). Hierarchical logistic regression models for clustered binary outcomes in studies of IVF-ET. *Fertil. Steril.* **73**(03), 575–581. (Cited on p. 117.)

Ingold, C. T., and Hadland, S. A. (1959). The ballistics of sordaria. *New Phytol.* **58**, 46–57. (Cited on pp. 98 and 100.)

Jackson, M. (2008). Content analysis, in *Research Methods for Health and Social Care*, pages 78–91. Palgrave Macmillan. (Cited on p. 74.)

Jeffreys, H. (1939). *The Theory of Probability.* Third edition published in 1998. Oxford University Press. (Cited on p. 144.)

Johnson, B. D. (2006). The multilevel context of criminal sentencing: integrating judge- and county-level influences. *Criminology* **44**, 259–298. (Cited on p. 116.)

Jöreskog, K. G., and Goldberger, A. S. (1975). Estimation of a model with multiple indicators and multiple causes of a single latent variable. *J. Am. Statist. Assoc.* **70**, 631–639. (Cited on p. 74.)

Kalbfleisch, J. D., and Prentice, R. L. (2002). *Statistical Analysis of Failure Time Data.* Second edition. Wiley. (Cited on p. 139.)

Knapper, C. M., Roderick, J., Smith, J., Temple, M., and Birley, H. D. L. (2008). Investigation of an HIV transmission cluster centred in South Wales. *Sex. Transm. Infect.* **84**, 377–380. (Cited on p. 82.)

Knudson, A. G. (2001). Two genetic hits (more or less) to cancer. *Nat. Rev. Cancer* **1**, 157–162. (Cited on p. 99.)

Koga, S., Maeda, T., and Kaneyasu, N. (2008). Source contributions to black carbon mass fractions in aerosol particles over the northwestern Pacific. *Atmos. Environ.* **42**, 800–814. (Cited on p. 80.)

Koukounari, A., Fenwick, A., Whawell, S., *et al.* (2006). Morbidity indicators of *Schistosoma mansoni*: relationship between infection and anemia in Ugandan schoolchildren before and after praziquantel and albendazole chemotherapy. *Am. J. Trop. Med. Hyg.* **75**, 278–286. (Cited on p. 115.)

Kremer, J. M., Rynes, R. I., and Bartholomew, L. E. (1987). Severe flare of rheumatoid arthritis after discontinuation of long-term methotrexate therapy. Double-blind study. *Am. J. Med.* **82**, 781–786. (Cited on p. 85.)

Krewski, D., Burnett, R. T., Goldberg, M. S., *et al.* (2000). Reanalysis of the Harvard Six Cities Study and the American Cancer Society Study of particulate air pollution and mortality. Special report, Health Effects Institute. (Cited on p. 76.)

Krewski, D., Burnett, R. T., Goldberg, M. S., *et al.* (2003). Overview of the Reanalysis of the Harvard Six Cities Study and the American Cancer Society Study of Particulate Air Pollution and Mortality. *J. Toxicol. Environ. Health A* **66**, 1507–1551. (Cited on p. 76.)

Lehmann, E. L. (1990). Model specification. *Statist. Sci.* **5**, 160–168. (Cited on p. 117.)

Li, B., Nychka, D. W., and Ammann, C. M. (2010). The value of multiproxy reconstruction of past climate (with discussion). *J. Am. Statist. Assoc.* **105**, 883–911. (Cited on p. 2.)

Lyons, L. (2008). Open statistical issues in particle physics. *Ann. Appl. Statist.* **2**, 887–915. (Cited on p. 88.)

Maconochie, N., and Roman, E. (1997). Sex ratios: are there natural variations within the human population? *Br. J. Obstet. Gynaecol.* **104**, 1050–1053. (Cited on p. 145.)

Manski, C. (1993). Identification problems in the social sciences, in *Sociological Methodology*, pages 1–56.. Jossey-Bass. (Cited on p. 87.)

Martel, M., Negrín, M. A., and Vázquez-Polo., F. J. (2009). Eliciting expert opinion for cost-effectiveness analysis: a flexible family of prior distributions. *SORT* **33**, 193–212. (Cited on p. 143.)

Mayo, D. G. (1996). *Error and the Growth of Experimental Knowledge*. University of Chicago Press. (Cited on p. 13.)

McCarthy, H. D., Ellis, S. M., and Cole, T. J. (2003). Central overweight and obesity in British youth aged 11–16 years: cross sectional surveys of waist circumference. *Br. Med. J.* **326**, 624. (Cited on p. 47.)

McCarthy, M. A., and Parris, K. M. (2004). Clarifying the effect of toe clipping on frogs with Bayesian statistics. *J. Appl. Ecol.* **41**, 780–786. (Cited on p. 36.)

McNamee, R. (2002). Optimal designs of two-stage studies for estimation of sensitivity, specificity and positive predictive value. *Stat. Med.* **21**, 3609–3625. (Cited on p. 80.)

McShane, B. B., and Wyner, A. J. (2011). A statistical analysis of multiple temperature proxies: Are reconstructions of surface temperatures over the last 1000 years reliable? (with discussion). *Ann. Appl. Statist.* **5**, 5–123. (Cited on p. 2.)

Meinshausen, N., and Bühlmann, N. (2010). Stability selection (with discussion). *J. R. Statist. Soc. B* **72**, 417–474. (Cited on p. 139.)

Merton, R. K., and Lazarsfeld, P. F. (1950). *Continuities in Social Research: Studies in the Scope and Method of "The American Soldier"*. The Free Press. (Cited on p. 74.)

Messer, L. C., Oakes, J. M., and Mason, S. (2010). Effects of socioeconomic and racial residential segregation on preterm birth: a cautionary tale of structural confounding. *Am. J. Epidemiol.* **171**, 664–673. (Cited on p. 87.)

Michaels, P. J. (2008). Evidence for publication bias concerning global warming in *Science* and *Nature*. *Energy Environ.* **19**, 287–301. (Cited on p. 181.)

Moore, R. J., Jones, D. A., Cox, D. R., and Isham, V. S. (2000). Design of the HYREX raingauge network. *Hydrol. Earth Syst. Sci.* **4**, 521–530. (Cited on p. 35.)

Office for National Statistics (2010). Consumer price indices: a brief guide. (Cited on p. 64.)

O'Mahony, R., Richards, A., Deighton, C., and Scott, D. (2010). Withdrawal of disease-modifying antirheumatic drugs in patients with rheumatoid arthritis: a systematic review and meta-analysis. *Ann. Rheum. Dis.* **69**, 1823–1826. (Cited on p. 85.)

Pepe, M. S., and Janes, H. (2007). Insights into latent class analysis of diagnostic test performance. *Biostatistics* **8**, 474–484. (Cited on p. 68.)

Powell, S. G., Baker, K. R., and Lawson, B. (2009). Errors in operational spreadsheets. *J. Organ. End User Comput.* **21**(3), 24–36. (Cited on p. 78.)

Prentice, R. L. (1989). Surrogate endpoints in clinical trials: definition and operational criteria. *Stat. Med.* **8**, 431–440. (Cited on p. 60.)

R Development Core Team (2007). R: a language and environment for statistical computing. The R Foundation for Statistical Computing. (Cited on p. 110.)

Raghunathan, T. E., and Grizzle, J. E. (1995). A split questionnaire survey design. *J. Am. Statist. Assoc.* **90**, 54–63. (Cited on p. 80.)

Rathbun, S. L. (2006). Spatial prediction with left-censored observations. *J. Agr. Biol. Environ. Stat.* **11**, 317–336. (Cited on p. 62.)

Reeves, G. K., Cox, D. R., Darby, S. C., and Whitley, E. (1998). Some aspects of measurement error in explanatory variables for continuous and binary regression models. *Stat. Med.* **17**, 2157–2177. (Cited on pp. 72 and 74.)

Riley, S., Fraser, C., Donnelly, C. A., *et al.* (2003). Transmission dynamics of the etiological agent of SARS in Hong Kong: impact of public health interventions. *Science* **300**, 1961–1966. (Cited on p. 92.)

Ripley, B. D. (1987). *Stochastic Simulation*. Wiley. (Cited on p. 158.)

Robert, C. P., and Casella, G. (2004). *Monte Carlo Statistical Methods*. Springer. (Cited on p. 158.)

Roethlisberger, F. J., and Dickson, W. J. (1939). *Management and the Worker*. Harvard University Press. (Cited on p. 55.)

Rosenthal, R. (1979). The file drawer problem and tolerance for null results. *Psychol. Bull.* **86**, 638–641. (Cited on p. 181.)

Ross, G. (1990). *Nonlinear Estimation*. Springer. (Cited on p. 139.)

Salcuni, S., Di Riso, D., Mazzesch, C., and Lis, A. (2009). Children's fears: a survey of Italian children ages 6 to 10 years. *Psychol. Rep.* **104**(06), 971–988. (Cited on p. 63.)

Samph, T. (1976). Observer effects on teacher verbal classroom behaviour. *J. Educ. Psychol.* **68**(12), 736–741. (Cited on p. 54.)

Scheffé, H. (1959). *The Analysis of Variance*. John Wiley & Sons. (Reprinted by Wiley-Interscience in 1999). (Cited on pp. 52 and 158.)

Schweder, T., and Spjøtvoll, E. (1982). Plots of *p*-values to evaluate many tests simultaneously. *Biometrika* **693**, 493–502. (Cited on p. 158.)

Shahian, D. M., Normand, S. L., Torchiana, D. F., *et al.* (2001). Cardiac surgery report cards: comprehensive review and statistical critique. *Ann. Thorac. Surg.* **72**, 2155–2168. (Cited on p. 55.)

Shapter, T. (1849). *The History of the Cholera in Exeter in 1832*. John Churchill. (Cited on p. 82.)

Shiboski, S., Rosenblum, M., and Jewell, N. P. (2010). The impact of secondary condom interventions on the interpretation of results from HIV prevention trials. *Statist. Commun. Infect. Dis.* **2**, 2. (Cited on p. 40.)

Snedecor, G. W., and Cochran, W. G. (1956). *Statistical Methods Applied to Experiments in Agriculture and Biology*. Fifth edition. Iowa State University Press. (Cited on pp. 52 and 183.)

Snijders, T. A. B., and Bosker, R. J. (2011). *Multilevel Analysis*. Second edition. Sage. (Cited on p. 183.)

Snow, J. (1855). *On the Mode of Communication of Cholera*. John Churchill. (Cited on pp. 82 and 83.)

Solow, R. M. (1970). *Growth Theory: An Exposition*. Paperback edn. Oxford University Press. (Cited on p. 109.)

Stamler, J. (1997). The INTERSALT study: background, methods, findings, and implications. *Am. J. Clin. Nutr.* **65**, 626S–642S. (Cited on p. 72.)

Sterling, T. D. (1959). Publication decisions and their possible effects on inferences drawn from tests of significance – or vice versa. *J. Am. Statist. Assoc.* **54**, 30–34. (Cited on p. 181.)

Storey, J. D. (2002). A direct approach to false discovery rates. *J. R. Statist. Soc. B* **64**, 479–498. (Cited on p. 158.)

Stubbendick, A. L., and Ibrahim, J. G. (2003). Maximum likelihood methods for nonignorable missing responses and covariates in random effects models. *Biometrics* **59**(12), 1140–1150. (Cited on p. 81.)

Sullivan, M. J., Adams, H., and Sullivan, M. E. (2004). Communicative dimensions of pain catastrophizing: social cueing effects on pain behaviour and coping. *Pain* **107**, 220–226. (Cited on p. 54.)

ten Wolde, S, Breedveld, F. C., Hermans, J., *et al.* (1996). Randomised placebo-controlled study of stopping second-line drugs in rheumatoid arthritis. *Lancet* **347**, 347–352. (Cited on p. 85.)

Thompson, M. E. (1997). *Theory of Sample Surveys.* Chapman and Hall. (Cited on p. 51.)

Thompson, S. K. (2002). *Sampling.* Second edition. John Wiley & Sons (First edition published in 1992). (Cited on p. 51.)

Tippett, L. H. C. (1927). *Random Sampling Numbers.* Cambridge University Press. (Cited on p. 158.)

Todd, T. J. (1831). *The Book of Analysis, or, a New Method of Experience: Whereby the Induction of the Novum Organon is Made Easy of Application to Medicine, Physiology, Meteorology, and Natural History: To Statistics, Political Economy, Metaphysics, and the More Complex Departments of Knowledge.* Murray. (Cited on p. 86.)

Toscas, P. J. (2010). Spatial modelling of left censored water quality data. *Environmetrics* **21**, 632–644. (Cited on p. 62.)

Van der Leeden, H., Dijkmans, B. A., Hermans, J., and Cats, A. (1986). A double-blind study on the effect of discontinuation of gold therapy in patients with rheumatoid arthritis. *Clin. Rheumatol.* **5**, 56–61. (Cited on p. 85.)

Vandenbroucke, J. P., and Pardoel, V. P. A. M. (1989). An autopsy of epidemiological methods: the case of 'poppers' in the early epidemic of the acquired immunodeficiency syndrome (AIDS). *Am. J. Epidemiol.* **129**, 455–457. (Cited on p. 71.)

Vogel, R., Crick, R. P., Newson, R. B., Shipley, M., Blackmore, H., and Bulpitt, C. J. (1990). Association between intraocular pressure and loss of visual field in chronic simple glaucoma. *Br. J. Ophthalmol.* **74**, 3–6. (Cited on p. 61.)

Walshe, W. H. (1841a). Tabular analysis of the symptoms observed by M. Louis, in 134 cases of the continued fever of Paris (Affection Typhoïde). *Provincial Med. Surg. J.,* **2**(5), 87–88. (Cited on p. 86.)

Walshe, W. H. (1841b). Tabular analysis of the symptoms observed by M. Louis, in 134 cases of the continued fever of Paris (Affection Typhoïde). *Provincial Med. Surg. J.,* **2**(5), 107–108. (Cited on p. 86.)

Walshe, W. H. (1841c). Tabular analysis of the symptoms observed by M. Louis, in 134 cases of the continued fever of Paris (Affection Typhoïde). *Provincial Med. Surg. J.,* **2**(5), 131–133. (Cited on p. 86.)

Webb, E., and Houlston, R. (2009). Association studies, in *Statistics and Informatics in Molecular Cancer Research*, pp. 1–24. Oxford University Press. (Cited on p. 154.)

Wellcome Trust Case Control Consortium. (2007). Genome-wide association study of 14 000 cases of seven common diseases and 3000 shared controls. *Nature* **447**, 661–678. (Cited on p. 152.)

Wilansky-Traynor, P., and Lobel, T. E. (2008). Differential effects of an adult observer's presence on sex-typed play behavior: a comparison between gender-schematic and gender-aschematic preschool children. *Arch. Sex. Behav.* **37**, 548–557. (Cited on p. 54.)

Wilson, E. B. (1952). *An Introduction to Scientific Research*. McGraw-Hill (Reprinted in 1990 by Dover.) (Cited on p. 13.)

Woodroffe, R., Donnelly, C. A., Cox, D. R., *et al.* (2006). Effects of culling on badger *Meles meles* spatial organization: implications for the control of bovine tuberculosis. *J. Appl. Ecol.* **43**, 1–10. (Cited on p. 17.)

Writing group for Women's Health Initiative Investigators (2002). Risks and benefits of estrogen plus progestin in healthy postmenopausal women. *J. Am. Med. Assoc.* **288**, 321–333. (Cited on p. 7.)

Wu, S. X., and Banzhaf, W. (2010). The use of computational intelligence in intrusion detection systems: a review. *Appl. Soft Comput.* **10**, 1–35. (Cited on p. 79.)

Yates, F. (1952). Principles governing the amount of experimentation required in development work. *Nature* **170**, 138–140. (Cited on p. 28.)

Index

Lightning Source UK Ltd.
Milton Keynes UK
UKOW06n1531230715

255721UK00001B/1/P

9 781107 013599